NEW APPROACHES IN CELL BIOLOGY

Symposium on New Approaches in Cell Biology, Imperial College of Science and Technology, 1958

NEW APPROACHES IN CELL BIOLOGY

Proceedings of a Symposium held at Imperial College, London, July 1958

Edited by

P. M. B. WALKER

*Royal Society Research Fellow
Department of Zoology
University of Edinburgh*

1960

ACADEMIC PRESS

LONDON AND NEW YORK

ACADEMIC PRESS INC. (LONDON) LTD.
17 OLD QUEEN STREET
LONDON, S.W.1.

U.S. Edition, Published by

ACADEMIC PRESS INC.
111 FIFTH AVENUE
NEW YORK 3, NEW YORK

COPYRIGHT ©, 1960, BY ACADEMIC PRESS INC. (LONDON) LTD.

Library of Congress Catalog Card Number: 60-8057

Printed in The Netherlands by
JOH. ENSCHEDÉ EN ZONEN GRAFISCHE INRICHTING N.V.

PREFACE

The Fifteenth International Congress of Zoology, which was held in London, included among its many sections this symposium on "New Approaches in Cell Biology". The latter thus represents a field of research which, while differing from much that was included in the remainder of the Congress, did not depart from the great tradition of cytology to which so many zoologists have made notable contributions.

It is nevertheless true that there is an increasing gulf dividing those who study the animal and its organs as a functional entity, and those who are attempting to investigate the basic units from which organs and animals are made. This is largely due to the great changes that have occurred in cell biology in the last ten years: increasingly the cell biologist has come to depend on the physicist for extending the precision of his observation and measurement, on the biochemist for his understanding of his chemical procedures, and on molecular biology and the genetics of micro-organisms for his basic concepts and assumptions. Since many, but not all, university departments in pure biology fail to provide either the facilities or the wide range of intellectual disciplines on which cell research depends, it is frequently undertaken in other institutions, and is often isolated from the broader aspects of biology.

This situation is, paradoxically enough, probably more dangerous to the cell biologist. Not only is he isolated from the source of new recruits to the study of his subject, but he may also be too narrowly constricted in his choice of material, and, even more important, he may well be ill-equipped for the next step in the development of cell biology—the study of the control of cell growth and differentiation. On the other hand, the more classical biologist may be unaware of the many new methods which have been used successfully in cellular research, and which may be equally useful in other fields. This, fortunately, can be more easily rectified than the difficulties of the cell biologist.

This volume is one attempt to provide a survey of some fruitful lines of research, for each contribution is valuable not only for the special results it discusses but as an example of a method which may be helpful in the solution of quite unrelated problems. The reader will be gently led, if he allows, from the opening papers, which he may properly consider to be within the provenance of classical zoology, through problems involving the biochemical approach, to those methods for studying the chemistry of single cells, which are made possible by the use of elegant physical techniques. He will also, I hope, join in thanking the authors for all their hard work in making this volume possible.

P. M. B. WALKER

LIST OF CONTENTS

	Page
NUCLEAR TRANSFER OF EMBRYONIC CELLS OF THE AMPHIBIA J. A. MOORE, *Department of Zoology, Columbia University, New York, U.S.A.*	1
CELLULAR INHERITANCE AS STUDIED BY NUCLEAR TRANSFER IN AMOEBAE J. F. DANIELLI, *Department of Zoology, King's College, London, England*	15
LAMPBRUSH CHROMOSOMES H. G. CALLAN and L. LLOYD, *Department of Natural History, The University, St. Andrews, Scotland*	23
THE MORPHOLOGY OF DEVELOPING SYSTEMS AT THE ULTRAMICROSCOPICAL LEVEL C. H. WADDINGTON, *Institute of Animal Genetics, Edinburgh, Scotland*	47
THE ORIGIN OF THE NUCLEUS AFTER MITOTIC CELL DIVISION J. BOSS, *Department of Physiology, University of Bristol, England*	59
LABELLED ANTIBODIES IN THE STUDY OF DIFFERENTIATION R. M. CLAYTON, *Institute of Animal Genetics, Edinburgh, Scotland*	67
A BIOCHEMICAL APPROACH TO CELL MORPHOLOGY W. S. VINCENT, *Department of Anatomy, Upstate Medical Center, State University of New York, Syracuse, New York, U.S.A.*	89
PAPER CHROMATOGRAPHY IN RELATION TO GENETICS AND TAXONOMY A. A. BUZZATI-TRAVERSO, *Istituto di Genetica, Università di Pavia, Pavia, Italy, and Scripps Institution of Oceanography, University of California, La Jolla, California, U.S.A.*	95
THE TRANSFER MECHANISMS IN ACTIVE TRANSPORT W. D. STEIN, *Department of Colloid Science, Cambridge, England*	125
THE MATCHING OF DRUGS TO TUMOURS P. HEBBORN, *Department of Zoology, King's College, London, England*	139
THE CYTOCHEMISTRY OF PROTEINS E. A. BARNARD, *Department of Zoology, King's College, London, England*	145

LIST OF CONTENTS

	Page
CYTOCHEMISTRY OF NUCLEIC ACIDS	155
C. LEUCHTENBERGER, *Children's Cancer Research Foundation, Boston, Massachusetts, U.S.A.*	
THE INTERFERENCE MICROSCOPE AS A CELL BALANCE	173
A. J. HALE, *Imperial Cancer Research Fund, London, England*	
FLYING SPOT MICROSCOPY	187
W. K. TAYLOR, *Department of Anatomy, University College, London, England*	
AUTHOR INDEX	199
SUBJECT INDEX	207

NUCLEAR TRANSFER OF EMBRYONIC CELLS OF THE AMPHIBIA

J. A. Moore

Department of Zoology, Columbia University, New York

One of the standard methods in experimental science for obtaining an understanding of the interrelations between two phenomena is to vary one while keeping the other constant. Unfortunately, this method can only be applied in a limited way when studying the interrelation between the nucleus and the cytoplasm. In most organisms the only practical way to vary the nucleus or the cytoplasm is through the selection of the gametes that are used to form a zygote.

Thus the effects of a nucleus and of its genes are measured on a time scale that has the generation as its unit. The existence and activitities of the genes that are assembled at fertilization are revealed by the developing phenotype of the organism of which they form a part.

Important information can be provided for many biological problems by mixing nuclei and cytoplasms at times other that at fertilization. One of the pressing problems for which such information is desired is the functional state of nuclei and cytoplasms during cellular differentiation. For example, when the paths of differentiation of the ectodermal cells diverge to produce the specialized cells of the lens, retina, brain, or epidermis, is there a concomittant change in the nucleus, or in the cytoplasm, or in both? Perhaps the most generally held view is that the nucleus maintains its full genetic potentialities in all cells and throughout the life cycle of the individual. Admittedly the evidence for this belief is inconclusive for all cells other than those that give rise to the gametes, but many observations, especially those based on regeneration, suggest that this view is correct.

The prime difficulty in studying this problem is the inability to use somatic cells or combinations of their nuclei and cytoplasms as the progenitors of new individuals. Even if we transfer a nucleus from one cell to another, it is not enough merely to replace the old nucleus. The cell with the new nucleus has to give rise to other cells or develop into something which can express the interrelations of the newly produced nuclear-cytoplasmic combination.

There are three likely ways of carrying out such an experiment and success has been achieved in two of them.

1. *Nuclear Transfer in Unicellular Organisms.* This was first achieved in amoeba by Commandon and de Fonbrune [5] by the following method: With a blunt needle the nucleus is pushed out of one amoeba and this enucleated individual is placed adjacent to one with a nucleus; this nucleus is then pushed from its own cell into the enucleated individual. In successful experiments the cell that has received the nucleus will reproduce by mitosis to form a clone of new individuals. Danielli and his associates [6–8; 19–21] have transferred the nucleus of one species of amoeba to the enucleated cytoplasm of another species. The general conclusion from these experiments was that the cytoplasm exerts a profound and continuing influence on the nucleus, since after a number of years, the chimera was identical neither with the nuclear nor the cytoplasmic parent but had its own individual characteristics which seemed to result from the co-adaptation of the nucleus of one species and the cytoplasm of the other. These results are discussed by Professor Danielli in this Symposium.

2. *Nuclear Transfer in Tissue-Culture Cells.* A similar experiment could be performed with two different cell types growing in tissue culture. If a cell composed of the cytoplasm of one type and the nucleus of another type would give rise to a clone of cells, it would be possible to study the influences of the nucleus and cytoplasm. Experiments of this sort have been attempted in several laboratories, but so far without success.

3. *Nuclear Transfer in Embryonic Cells.* A third method has been developed by Briggs and King [1–4; 13–16] to study the question of changes in the nucleus during the course of embryonic development in frog eggs. It is with this method and its results that my paper is concerned.

I. Methods of Transferring Nuclei in Amphibia

The two main components needed in an ideal experiment involving nuclear transfer are first, a cell without a nucleus and second, a nucleus devoid of cytoplasmic contaminants. The first of these is available, but the second is not.

The method of obtaining a cell without a nucleus is a slight modification of the technique developed by Porter [26]. When the ovum of *Rana pipiens* leaves the female its nucleus is situated immediately below the surface at the animal pole and is in metaphase of the second meiotic division. If the ovum is fertilized, there are changes in the surface immediately over the maternal nucleus. A tiny pit forms and under proper illumination this appears as a minute black spot. The entrance of the sperm also stimulates the egg to rotate and the animal pole comes to be uppermost; this makes it possible to observe the site of the maternal pronucleus. It is a simple procedure to remove the maternal nucleus by pushing a glass needle laterally beneath the black spot and then raising the needle. A small exovate containing the maternal nucleus is formed. This operation is performed freehand and, with practice, it is 99 to 100% successful. No developmental disturbances result from the operation *per se*. The embryo is a haploid since it contains only the chromosomes brought in by the sperm.

Obviously this method does not result in an enucleated cell. One nucleus is removed, to be sure, but the sperm brings in another. The sperm is not essential as a stimulus to rotation and meiosis, however. The rotation of the ovum and the formation of the black spot occur just as readily if the ovum is activated by being pricked with a clean glass needle.

The enucleated host cells, therefore, are prepared as follows: Several hundred ova are stripped from the female. Each ovum is pricked with a glass needle and then the maternal nucleus is removed. The length of time during which it is possible to enucleate the ova is 15 to 25 minutes—enough to perform about 100 operations.

The cells providing the donor nuclei in the first experiments performed by Briggs and King were from the inner side of the blastocoel roof of a late blastula or early gastrula. Such a cell was drawn into a micropipette having a smaller diameter than the cell. As a result the cell membrane was broken but the nucleus remained intact. The cell was then injected into a previously enucleated ovum. This operation can be performed freehand but it is accomplished more easily with a simple micromanipulator. As was mentioned earlier, the ideal experiment of transferring a nucleus entirely devoid of cytoplasm has not been possible up to the present since the solution in which the transfers are made damages the nucleus unless it is surrounded by a protective coat of cytoplasm. In actual practice, therefore, a ruptured cell is transferred to the enucleated ovum.

About one-third of the enucleated ova that were injected cleaved and formed normal blastulae; of these, about two-thirds went on to form normal or slightly abnormal larvae.

There were elaborate controls for these experiments and they showed that it was the injected nucleus that gave rise to the nuclei of the embryo. The possibility that the maternal nucleus had been retained and that parthenogenesis was involved was excluded. Although the method gave far less than complete success, enough embryos developed so that the method could be used to compare the nuclei from donors of different ages.

In the first experiments nuclei of cells from the inner side of the blastocoel roof were transferred. These cells form ectodermal structures in the course of development. Using the conventional techniques of experimental embryology, it has been shown that in a late blastula these cells are "undetermined"; that is, if they are transplanted to other parts of the embryo they can form different structures. The nuclear transfer experiments substantiated this point of view.

At the end of gastrulation, however, different results are obtained when parts of the embryo are transplanted or explanted. The portion of the ectoderm which is destined to form the neural tube will do so even when removed from its normal position in the embryo. It is said to be "determined", even though at this time there is no visible histological evidence for its newly acquired state.

Briggs and King had developed their method largely to study the nuclei of such determined cells. In 1954 they presented the results of transfer of nuclei from the chorda-mesoderm of the archenteron roof and the ectoderm of the overlying presumptive medullary plate regions of a late gastrula [14]. At this stage the chorda-mesoderm is determined and, if explanted, will differentiate into mesodermal and other structures. The presumptive medullary plate cells are also determined and, if explanted, will form neural tissue. When broken cells, with nucleus intact and cytoplasm somewhat dispersed, from either of the regions, were transferred to enucleated ova, some embryos were obtained. Of those that developed normally in the pre-gastrula stages, and therefore were not obviously injured by the experiment, 35% of those containing a nucleus from the chorda-mesoderm cell formed larvae. Of those containing a nucleus from a presumptive medullary plate cell, 36% formed larvae. Parallel experiments involving the transfer of nuclei of undetermined cells from the presumptive epidermis region of an early gastrula gave larvae in 59% of the cases.

These experiments showed that neither the intact nucleus nor the disrupted cytoplasm of the transferred cell had become differentiated in some irreversible manner that would alter the course of development when transferred to an enucleated ovum. It is necessary to emphasize that no *irreversible* change had occurred. Briggs and King have been careful to point out from the very first that their method does not detect reversible changes. It is entirely possible, and I believe even probable, that a nucleus of a chorda-mesoderm cell of a late gastrula is different from one in a presumptive medullary plate cell. The experiments of Briggs and King showed, however, that, whatever the state of the nucleus at this stage, it is capable of participating in entirely normal development when transferred to an enucleated ovum.

The large difference in the per cent forming larvae, 35 and 36 for the late gastrula nuclei compared to 59 for the early gastrula nuclei, was thought to be a measure of the relative susceptibility of the nuclei to injury. It is essential that a nucleus be coated with a layer of cytoplasm to protect it from contact with the artificial medium (Niu-Twitty solution) in which the transfers are made. Since the early gastrula cells were larger and have more cytoplasm, their nuclei are afforded greater protection.

The hypothesis that the decrease in effectiveness of the late gastrula nuclei was due to their greater susceptibility to injury was not entirely satisfactory. The phenomenon could also be interpreted as a consequence of changes within the nuclei themselves. King and Briggs [15] undertook two sorts of experiments in an effort to distinguish between the hypotheses.

First, they improved their technique to a point where more than twice as many of the transfer embryos were developing normally to the late blastula stage, but still they observed a low percentage of normal development after transfers of older nuclei. Of the embryos containing a nucleus of a chorda-mesoderm cell of a late gastrula and developing to the late blastula stage, only 11% were able to continue and form larvae. Control experiments with the undetermined presumptive ectodermal cells of an early gastrula, gave larvae in 86% of the cases.

Second, they conducted experiments on the endodermal cells of a late gastrula, which are known from transplantation and explantation to be determined. These cells are large and hence have a more effective protective blanket of cytoplasm. The development of embryos containing a nucleus from such a cell was most interesting. First, only 19% of those forming normal blastulae were able to develop into larvae. This percentage should be compared with the value of 86% for the nuclei of the presumptive ectodermal cells of the early gastrula. The evidence was becoming stronger that the nuclei were undergoing some change between the beginning and end of gastrulation. The quantitative evidence was suggestive, but even more convincing evidence was obtained by a study of the most typical (50% of the embryos were in the class) of the embryos, which ceased developing in neurula and postneurula stages. "They grew to a length of about 6 mm. and displayed a combination of deficiencies which we have not observed in other embryos. This syndrome is characterized by a loss of the integrity of the epidermis, which becomes thickened in some places and thin or absent in others. Internally, the notochord is well developed, somites are large but abnormal in forms and the gut is developed as well as the general condition of the embryo allows. All ectodermal derivatives, however, are very poorly developed and display degenerative nuclear changes. It should be added that these effects on development are not observed when endoderm cytoplasm alone is injected into either enucleated or normally nucleated eggs."

These experiments were greatly extended [4] by using endodermal nuclei from not only a late gastrula but also the neurula and tailbud stages. It was found that with increasing age the endodermal nuclei showed a progressive decrease in their ability to provide for normal cleavage when transferred to enucleated ova. In addition, there was a tendency for the onset of the first cleavage to be delayed for a time approximating the normal intercleavage period. Nearly all of these embryos with retarded first cleavage formed tetraploid embryos.

These experiments confirmed the earlier ones with the endodermal nuclei of late gastrula. Only 20% of the embryos forming normal blastulae were able to continue their development to form normal larvae, whereas controls with presumptive ectodermal cells of the early gastrula gave 85% larvae. The corresponding figures for embryos containing nuclei from neurula and tailbud stages were 6 and 0%. Most of the embryos were abnormal. The chief defects were in the structures derived from the ectoderm: the epidermis was thin or abnormal; the central nervous system was deficient; the neural crest derivatives were lacking or poorly developed; nasal pits were always absent; eyes and otic vesicles were absent in most embryos. In addition, pycnotic nuclei were common. The mesodermal and endodermal derivatives, on the other hand, were more normal: they contained few or no degenerating nuclei.

Throughout this discussion of the experiments of Briggs and King the emphasis has been placed on the effect of the nucleus. The cytoplasm of the injected cell has been thought to serve no role other than to protect the nucleus from contact with the Niu-Twitty solution in which the operations are performed. There was no need to be concerned about the role of the cytoplasm so long as the nuclei gave no evidence of being differentiated—as was the case with the transfer of the early gastrula presumptive ectoderm cells. Once positive evidence of altered development was obtained—as was the case with the transfer of late gastrula and older endoderm cells—it became necessary to determine the relative contributions of nucleus and cytoplasm. Evidence that the cytoplasm of the transferred cell has no influence on the type of development is based on two main sorts of evidence:

1. The volume of transferred cytoplasm is small relative to that of the enucleated ovum. Thus the volume of cytoplasm in the endodermal cells is 1/40,000 or less of that in the enucleated ovum.

2. When normally fertilized ova are injected with cytoplasm of endodermal cells taken from late gastrula, mid-neurula, and tailbud stage donors, they develop in a perfectly normal manner. They do not show any of the specific defects observed when endodermal nuclei with the surrounding cytoplasm of these stages are transferred to enucleated ova.

Thus it has been established, as well as is possible in the present stage of perfection of the nuclear-transfer technique, that it is the nucleus and not the cytoplasm of the transferred endodermal cell that gives evidence of differentiation. This conclusion must be a somewhat tentative one, however, as long as cytoplasm is transferred with the nucleus. The "ideal" experiment consisting of the transfer of the nucleus alone is not possible at the present time. It is even likely that it will always be impossible since an enucleated ovum injected with a clean nucleus would probably lack a division apparatus. An approximation to the "ideal" experiment could be carried out by transferring a clean nucleus to an enucleated cell fertilized with irradiated sperm (a dose sufficient to inactivate the nucleus but not that portion of the sperm that forms the division apparatus).

There is still another experiment of King and Briggs [16] that will be mentioned. It is indeed an elegant one and it provides the best evidence so far provided that nuclei of the endodermal cells of a late gastrula and older embryos are not capable of supporting normal development. The background observations are these: The embryos that originate from the transfers of late gastrula endoderm nuclei are quite diverse in their development. Of those that are normal in early stages some cease development as blastulae, a few as gastrulae, and many as the characteristically abnormal embryos with the defective ectodermal structures. A few form normal larvae. This would seem to indicate that there is considerable variation in the potentialities of the nuclei or that the results were due to injuries sustained by the nuclei when the transfers were made.

If we adopt the first alternative as a working hypothesis, then we assume that an embryo developing defective ectodermal derivatives has nuclei with limited potentialities, that is, the original nucleus transferred to the enucleated ovum was irreversibly determined. If this is so, then it should be possible to demonstrate the fact by transferring its nuclei to enucleated ova.

An elaborate experiment was necessary. The initial donor, which we will call A, was a normal late gastrula: a number of endodermal cells were taken from it and transferred to enucleated ova. We will identify the embryos so formed as B_1, B_2, B_3, etc. It is these embryos that will develop and produce the wide spectrum of results, which we are assuming is due to differences in their nuclei. When B_1, B_2, B_3, etc., are late blastulae, one is selected (we will call it B_1) as the donor and presumptive ectodermal cells from the blastocoel roof are transferred to enucleated ova. This second generation of embryos derived from B_1 nuclei will be called C_1, C_2, C_3, etc.

Of course, we never know what sort of embryo B_1 would have formed if we had not sacrificed it as a donor. We should be able to surmise this, however, from the development of its offspring C_1, C_2, C_3, etc. (The other members of the B series, namely B_2, B_3, etc., are kept and they show the expected range of defective embryos.)

It should be noted that B_1 was used when it was a blastula and the cells transferred were taken from the blastocoel roof. The earlier work of Briggs and King had showed that these nuclei are not differentiated; therefore, no further alteration of the original nucleus from embryo A during its replication in embryo B_1 would be expected.

Now to return to the offspring of B_1, namely C_1, C_2, C_3 etc. Most interestingly, these embryos showed very little morphological variation. In one experiment, for example, all the embryos were of the type with deficient ectodermal derivatives.

This method of making serial transfers of nuclei was continued for several more generations and the same type of defect occurred in each transfer generation. The uniformity of all the embryos of the clone was interpreted as a consequence of all having originated from a single differentiated endodermal nucleus that continued to show its limited potentialities even after repeated mitoses.

Many experiments of this sort were conducted. In some, all of the embryos were arrested in the gastrula stage, and in others as abnormal tailbud stages. A few experiments produced clones of nearly normal larvae.

The control experiments were started with a nucleus from a late blastula stage and hence was undifferentiated. The expected results were obtained: serial transfer gave clones of normal larvae.

The general conclusions that can be reached from the elegant experiments of Briggs and King are these:

1. The nuclei of the late blastula cells of *Rana pipiens* have not undergone any differentiation or, if they have, this differentiation is readily reversible upon transfer to an enucleated ovum.

2. By the end of the gastrula stage the nuclei of some of the endodermal cells have become irreversibly differentiated in the sense that, when transferred to an enucleated ovum, they cannot support normal development. The endodermal nuclei seem to become still more differentiated in older stages.

The nuclear-transfer technique has been used by Sambuichi to study the problem of overripe eggs [28].

II. Experiments with *Xenopus laevis*

Preliminary results have been reported by Fishberg, Gurdon, and Elsdale [10, 11, 12] on nuclear transfers in South African clawed frog, *Xenopus laevis*. Modifications of the Briggs and King technique were necessary for this species. The enucleation of the ovum presented a major problem until it was found that it was unnecessary to remove the maternal nucleus. When a nucleus is injected into a mature ovum of *Xenopus*, the maternal nucleus usually does not contribute to the nuclei of the embryo. It would have been difficult to prove this point were it not for the availability of a special nuclear marker. A single female *Xenopus* was found that had one nucleolus in diploid cells instead of the usual two [9]. From this single female a strain was established and it was shown that the single nucleolar condition is inherited in a typical Mendelian fashion. An individual with one nucleolus is heterozygous. When two such individuals are crossed, 25% of the offspring have one nucleolus, 50% have two, and 25% die as larvae. This last class is homozygous for the absence of a nucleolus.

With this convenient marker it was possible to transfer nuclei from an embryo with a single nucleolus to ova of a female with two nucleoli. It was found that the resulting embryo nearly always had a single nucleolus; this indicated that the maternal nucleus of the ovum was not participating.

In the work that has been reported by Fishberg and his associates, no conclusion has been reached regarding the possibility of nuclear differentiation; this is due, in part, to difficulties with the material itself. They have not found it possible to obtain eggs from *Xenopus* that give a consistently high percentage of fertilization. This difficulty has been surmounted in part by the performance of truly heroic numbers of transfers. In general, however, they do have less success with transfers of nuclei from older donors than from younger donors. Whether or not this is due to nuclear differentiation remains to be seen.

The positive results with nuclear transfers in *Xenopus* have been impressive. So far they have obtained 77 frogs from their nuclear transfer experiments. Some of these were obtained from donor embryos that, on the basis of Briggs' and King's findings with *Rana pipiens*, would be expected to have differentiated nuclei. Thus, nuclei from the mesoderm of a neural fold stage and from the endoderm of neural fold, tailbud, and muscular response stages all gave rise to adult frogs. The most advanced donor from which a normal individual was obtained was a stage 32 embryo: this would correspond to a stage 18 *Rana* embryo.

The general conclusion reached was that "some nuclei are capable of giving normal development very shortly before the organ of which they are part becomes functional; normal development was obtained from presumptive somite nuclei nine hours before the first muscular response."

This work of Fishberg and his associates is of great importance apart from the problems being studied. They have shown that *Xenopus* can be used for nuclear transfer experiments, a most important point since this species is perhaps the easiest of all frogs to be raised and bred in the laboratory.

III. Experiments with Urodele Embryos

The eggs of urodeles, which are so useful for many types of experimentation, are not well suited for nuclear transfer experiments. Several attempts [17, 18, 25, 29] have failed to produce embryos that develop beyond the blastula stage. If an amphibian embryo develops only to a late blastula stage, no useful conclusion can be reached regarding the state of its nucleus. It is well known [22] that a nucleus may be badly damaged, or quite unable to provide for normal development beyond gastrulation and yet the embryo can develop in a normal manner to a late blastula stage.

IV. Interspecific Nuclear Transfers

The nuclear transfer technique allows one to study a variety of problems concerned with the interaction of the nucleus of one species (or variety) and the cytoplasm of another. In some of their first experiments Briggs and King used nuclei of different species both to test their method of nuclear transfer as well as to obtain information about the nuclei themselves.

When an ovum of *Rana pipiens* is enucleated and fertilized with a *Rana catesbeiana* sperm, it develops normally, but only so far as the late blastula stage. When these haploid *catesbeiana* nuclei are then transferred to an enucleated *Rana pipiens* ovum, once again development is normal to the late blastula stage but the embryo is unable to gastrulate [1, 13]. Thus, repeated replication of the *catesbeiana* nucleus in *pipiens* cytoplasm does not increase its ability to form an embryo when transferred to another *pipiens* ovum.

The diploid hybrid made by the fertilization of a *pipiens* ovum with *catesbeiana* sperm develops normally to an early gastrula stage and then stops. When nuclei from a hybrid blastula are transferred to enucleated *pipiens* ova, the resulting embryo develops no further than the early gastrula stage [13]. Thus the hybrid nuclei retain their original potentialities even after numerous replication in the *pipiens* cytoplasm.

An experiment involving even more dissimilar parental types was tried [3]. Nuclei of a blastula of the newt, *Triturus pyrrhogaster*, were transferred to enucleated *Rana pipiens* ova. Cleavage was abnormal and generally restricted to the animal hemisphere. Most of the "cells" of the embryo were without a nucleus or chromatin.

These results with interspecific nuclear transfers might have been anticipated. In each case the two species are distantly related and, even in the case of the two frogs, the hybrids do not develop beyond an early gastrula stage. It is not surprising that the nucleus of one and the cytoplasm of the other fail to collaborate in a normal development. It seems likely that normal or near normal development would be possible only if the two species were closely related.

There has been only one report of nuclear transfers between closely related species. Sambuichi [27] transferred nuclei from a blastula of *Rana nigromaculata brevipoda* to enucleated ova of *Rana nigromaculata nigromaculata*.* One embryo in a total of 262 transfers reached the stage of metamorphosis and then died. The adult colour pattern was not fully developed, but in every respect that could be observed, the individual resembled *brevipoda*. So far as could be told the cytoplasm did not exert a specific effect.

I have performed nuclear transfers between *Rana pipiens* and *Rana sylvatica* [23]. These species are distantly related. Their hybrids develop only to the beginning of gastrulation. If nuclei from a *Rana sylvatica* blastula are transferred to enucleated *Rana pipiens* ova, the embryos develop normally to the end of the blastula stage. There is no further development, as could have been predicted from earlier experiments with androgenetic hybrids.

There are other problems, that do not have obvious answers, which can be studied with nuclear transfers between these species. If a nucleus of *Rana pipiens* is placed in an enucleated ovum of *Rana sylvatica*, it will undergo about 15 mitotic divisions by the end of the blastula stage when there are about 30,000 nuclei. Apart from the minute amount of cytoplasm introduced when the transfer was made, the original *pipiens* nucleus and its descendants would be replicated from the substrates present in the ovum. These substrates would have been formed under the influence of a *sylvatica* nucleus, which was removed, of course, before the *pipiens* nuclei was injected.

What is the nature of these *pipiens* nuclei which result from replications in the *sylvatica* cytoplasm? We can be sure that any one nucleus contains almost none of the original molecules of the transferred nucleus. Have the genes been able to duplicate themselves in a genetically exact fashion in spite of their sojourn in an alien biochemical

* Sambuichi believes that these two forms are subspecies. There is no doubt that they are very closely related for they form hybrids that reach the adult stage and are partially fertile. Furthermore, the androgenetic hybrids develop almost as well as the androgenetic controls. This is excellent evidence of their close relationship as is, of course, his success in obtaining the metamorphosing individual with the diploid nucleus of one form and the cytoplasm of the other. A better indication of taxonomic status, however, can be gained from a study of the behavior of two forms in the field if they are sympatric. Sambuichi notes that *brevipoda* and *nigromaculata* "differ (from) each other in quite numerous morphological characters besides in many ecological ones and there is nearly complete reproductive isolation between them." Furthermore, they "are sympatric in wide areas of Japan" and apparently show no intergradation. If two forms, which are not polymorphs (and obviously *brevipoda* and *nigromaculata* are not) are sympatric and maintain their distinctive characters, they demonstrate that they are different species. One could not ask for better evidence.

environment? An answer to this question can be obtained by transferring these nuclei back to enucleated *pipiens* ova, thus restoring the *pipiens* nuclei to their species-specific cytoplasm. Experiments of this type, involving the back transfer to *pipiens* cytoplasm of *pipiens* nuclei that had replicated in *sylvatica* cytoplasm, have been performed with diploid [23] and haploid [24] nuclei. The cleavage and blastula stages are passed normally and gastrulation begins, but this is generally the limit of developmental potentialities of the embryos. These results show that after the many replications in *sylvatica* cytoplasm the *pipiens* nuclei have lost their genetic potentialities.

The tentative hypothesis that was offered to explain these results is as follows: the mature frog's ovum contains large amounts of cytoplasmic substances that are either DNA or closely related molecules. The amounts have been estimated by different workers as sufficient to provide for all the nuclei of a blastula or even a neurula. Furthermore, there is little or no increase in the amount of DNA before gastrulation. These data suggest that during the cleavage and blastula stages the new chromosomes may be made from the DNA-like substances stored in the cytoplasm. In a normal embryo the *pipiens* chromosomes would be replicated from the cytoplasmic substances that were formed in the ovarian egg under the influence of *pipiens* chromosomes.

In the experimental embryos, however, the *pipiens* chromosomes would replicate from the *sylvatica* cytoplasmic DNA-like substances. The chromosomes that are formed are normal in morphology but the back transfer to *pipiens* cytoplasm shows they cannot function normally. If this explanation is correct, we must conclude that normal gene replication is a function of the whole cell and not solely a matter of the genes synthesizing exact copies of themselves from a passive soup of molecules in the cytoplasm. This is not a radical suggestion; if it seems so it is because biologists have repeated the dictum "genes are self-duplicating structures" for so long that we have forgotten that this dictum is meaningless in any operational sense.

Similar results have been obtained by Gurdon [11] using *Xenopus laevis* and *Rana temporaria*. A *Rana* nucleus was transferred to *Xenopus* and when the resulting embryo reached the blastula stage, nuclei were transferred back to enucleated *Rana* ova. These embryos developed to a late blastula stage but no further. Gurdon took nuclei from these embryos and transferred them a second time to enucleated *Rana* ova. In fact, the back transfers were done repeatedly and a most interesting phenomenon was noticed. For the first few transfers the embryos were unable to pass beyond the late blastula stage but usually by the fourth transfer some of the embryos formed abnormal gastrulae.

The hypothesis that is offered to account for the *Rana pipiens* and *Rana sylvatica* case can be extended to cover this additional situation. The essence of the hypothesis is that when chromosomes are in a foreign cytoplasm they use the existing cytoplasmic DNA-like molecules that are similar but not identical to those responsible for their normal structure. Thus when *Rana* chromosomes replicate in *Xenopus* cytoplasm, at least part of their structures will be abnormal—the precursors may be similar but not identical to those needed to form a normal *Rana* chromosome. When these "*Rana*" chromosomes, which could be thought of as analogues of normal *Rana* chromosomes, are returned to enucleated *Rana* ova, they would then replicate from the *Rana* DNA-like substances. After four back transfers, during which they would have replicated about 60 times, they seem to be more normal than after the first few transfers.

Unfortunately, *Rana temporaria* is not a very good species for nuclear transfer experimentation. Even when homospecific transfers are made, most of the best embryos stop developing as gastrulae. It is difficult, therefore, to evaluate the results of the transfers between *Rana* and *Xenopus*. However, the improvement that is noted after the fourth back transfer, namely that some of the embryos form gastrulae rather than stop as blastulae, may be more significant than it seems. If this is the limit of development of nuclear-transfer embryos in this species, then the improvement may be as much as can be measured. The reciprocal experiment, namely the transfer of *Xenopus* nuclei to *Rana* and then back to *Xenopus*, should be most interesting since in *Xenopus* the nuclear-transfer embryos may reach the adult stage.

V. Conclusions

The method of transferring nuclei of amphibian embryos, as described by Briggs and King in 1952, allows one to determine the all-important question of whether or not nuclei change in the course of embryonic differentiation. Briggs and King have presented evidence indicating that endodermal nuclei of *Rana pipiens* become progressively changed in post-gastrula stages. Fishberg and his associates have shown that at least some of the nuclei of larvae of *Xenopus laevis* are not irreversibly determined.

The method can also be used for interspecific transfers. If the parental species are closely related, the nuclear-transfer embryos seem to have the characteristics of the species from which the nucleus was derived. If the parents are remotely related, the nuclear-transfer embryos fail at the late blastula stage. There is evidence that the nuclei of such an embryo are profoundly altered during their r the foreign cytoplasm. Experiments of this nature should our understanding of some of the factors in gene rep meagre data available so far suggest that the control of g is at the level of the entire cell.

REFERENCES

1. BRIGGS, R. and KING, T. *Proc. Nat. Acad. Sci. U. S.* **38**, 455 (1952).
2. BRIGGS, R. and KING, T. *J. Exp. Zool.* **122**, 485 (1953).
3. BRIGGS, R. and KING, T. in "Biological Specificity and Growth." p. 207 Princeton University Press, Princeton, New Jersey (1955).
4. BRIGGS, R. and KING, T. *J. Morph.* **100**, 269 (1957).
5. COMMANDON, J. and DE FONBRUNE, P. *C. R. Soc. Biol. Paris* **130**, 740 (1939).
6. DANIELLI, J. F. *Sci. Amer.* **186**, 58 (1952).
7. DANIELLI, J. F. *Exp. Cell Res.* Suppl. **3**, 98 (1955).
8. DANIELLI, J. F., LORCH, I. J., ORD, M. J. and WILSON, E. G. *Nature, Lond.* **176**, 1114 (1955).
9. ELSDALE, T. R., FISHBERG, M. and SMITH, S. *Exp. Cell Res.* **14**, 642 (1958).
10. FISHBERG, M., GURDON, J. B. and ELSDALE, T. R. *Nature* **181**, 424 (1958).
11. FISHBERG, M., GURDON, J. B. and ELSDALE, T. R. *Exp. Cell Res.* Suppl. **6**, 161-178 (1958).
12. GURDON, J. B., ELSDALE, T. R. and FISHBERG, M. *Nature, Lond.* **182**, 64 (1958).
13. KING, T. J. and BRIGGS, R. *J. Exp. Zool.* **123**, 61 (1953).
14. KING, T. J. and BRIGGS, R. *J. Embryol. Exp. Morph.* **2**, 73 (1954).
15. KING, T. J. and BRIGGS, R. *Proc. Nat. Acad. Sci. U. S.* **41**, 321 (1955).
16. KING, T. J. and BRIGGS, R. *Cold Spr. Harb. Symp. Quant. Biol.* **21**, 271 (1956).
17. LEHMAN, H. G. *Biol. Bull.* **108**, 138 (1955).
18. LEHMAN, H. G. in "The Beginnings of Embryonic Development." Ed. Albert Tyler *et al.*, p. 201. Publication 48, Amer. Assoc. Adv. Sci.
19. LORCH, I. J. and DANIELLI, J. F. *Nature, Lond.* **166**, 329 (1950).
20. LORCH, I. J. and DANIELLI, J. F. *Quart. J. Micr. Sci.* **94**, 445 (1953).
21. LORCH, I. J. and DANIELLI, J. F. *Quart. J. Micr. Sci.* **94**, 461 (1953).
22. MOORE, J. A. *Advanc. Genet.* **7**, 139 (1955).
23. MOORE, J. A. *Exp. Cell Res.* **14**, 532 (1958).
24. MOORE, J. A. *Exp. Cell Res.* Suppl. **6**, 179-191 (1958).
25. PANTELOURIS, E. M. and JACOB, J. *Experientia* **14**, 99 (1958).
26. PORTER, K. R. *Biol. Bull.* **77**, 233 (1939).
27. SAMBUICHI, H. *J. Sci. Hiroshima Univ.* Ser. B, **17**, 33 (1957).
28. SAMBUICHI, H. *J. Sci. Hiroshima Univ.* Ser. B, **17**, 43 (1957).
29. WADDINGTON, C. H. and PANTELOURIS, E. M. *Nature, Lond.* **172**, 1050 (1953).

CELLULAR INHERITANCE AS STUDIED BY NUCLEAR TRANSFER IN AMOEBAE

J. F. Danielli

Department of Zoology, King's College, London

Problems Open to Investigation by Nuclear Transfer

During the past twelve years attempts have been made to use the technique of nuclear transfer to study a wide variety of problems. The use of this technique is valuable to determine whether a change, or process, is exclusively a function of either nucleus or cytoplasm. But it is equally useful in analyzing the relationships between nucleus and cytoplasm in phenomena in which both are concerned. Specific problems that can scarcely be solved without this technique are:

(1) The separate nuclear and cytoplasmic contributions to inheritance.
(2) The occurrence of changes that are irreversible, or can be reversed only with difficulty, in nucleus or cytoplasm during embryonic differentiation.
(3) The nature of inheritance in multiplication of differentiated cells, including the nature of the homeostatic processes which maintain a constant state of differentiation.
(4) As specific extensions of (3), the nature of the fundamental lesions in carcinogenesis and senescence at the cellular level.
(5) The cellular sites of damage by radiations and chemical agents.
(6) The nature of adaptation to chemical agents, and the manner of its inheritance.

The technique of nuclear transfer from cell to cell was first developed by Commandon and de Fonbrune [1], using *Amoeba sphaeronucleus* as the experimental material. This remarkable achievement was made possible only by the use of the de Fonbrune micromanipulator.

Unfortunately *A. sphaeronucleus* is not a satisfactory animal for culture under laboratory conditions, and to obtain satisfactory material recourse was had to *Amoeba proteus* and allied large mononucleate amoebae. With these large amoebae the transfer of nuclei is a little more difficult than with *sphaeronucleus*, as the large amoebae contain much granular and vesicular material which results in unfavourable optical conditions. But this is much more than compensated for by the ease with which standard animals can be obtained by simple culture methods.

Nature of results

Some of the results obtained by transplanting nuclei between strains and species will now be outlined. For this purpose amoebae are described in terms of the origin of their nucleus and cytoplasm: thus $_DP_n$, $_TD_c$ indicates an animal with a nucleus obtained from Dawson's strain of *Amoeba proteus* and a cytoplasm obtained from Sister Monica Taylor's strain of *Amoeba discoides*.

When studying the results of transferring nuclei from one species or strain of amoeba to another species or strain, the results obtained will fall into one of the following three categories (a) wholly nuclear inheritance (b) wholly cytoplasmic inheritance (c) dual control of inheritance.

Table 1

Value of averages of the maximum diameters of nuclei in various clones and heterotransfer clones of amoebae.

Type	$_{Tl}P_n, _{Tl}P_c$	$_TD_n, _{Tl}P_c$	$_{Tl}P_n, _TD_c$	$_TD_n, _TD_c$
Diameter in μ	45	44	38.6	38.2

Table 1 shows the results obtained by measuring the diameter of the nucleus of many members of four clones. Two of these clones were derived from the original strains *Amoeba proteus* and *Amoeba discoides* supplied to us by Sister Monica Taylor. The other two clones are those obtained by reciprocal exchange of nuclei between these two types of amoebae. All the strains obtained in this way have grown satisfactorily and vigorously. It will be seen that the heterotransfer containing proteus cytoplasm is closely similar to the strain containing both proteus nucleus and cytoplasm. Clearly the strain containing *discoides* cytoplasm approximates very closely to that typical of *Amoeba discoides*. This is not a transient relationship, but persists over many years, during which time the clones of amoebae are continually in active growth and division. Thus, as has been pointed out elsewhere, it is quite impossible to explain this approximation to the cytoplasmic type in terms of a mechanical carryover of macromolecules which were present in the cytoplasms of these animals at the time at which the nuclei were transferred.

On first inspection of the results it might be tempting to suppose that there was no significant effect of the nucleus exhibited in this experiment. This, however, is not so, for the differences between the parent strains and those strains containing only cytoplasm of that parent, though small, are persistent and remain of the same order of magnitude over many years. Thus in this case we have an example of control of a simple morphological character, in which the influence of the cytoplasm and the influence of the nucleus are both marked and persistent even years after the nuclei have been transferred. In this instance the effect of the cytoplasm is much more evident than is that of the nucleus.

Another character readily studied is that of the shape assumed by amoebae when moving by active amoeboid movement. The shape

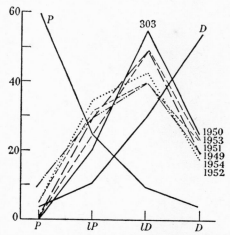

Fig. 1. Diagram showing the shapes assumed by various clones of amoebae when in active amoeboid movement. P indicates typical *proteus* shape; *IP*, more like *proteus* than like *discoides*; *ID*, more like *discoides* than like *proteus*; D, typical *discoides*. 303 is the clone P_nD_c (1), observed at intervals.

assumed by any strain when moving under standard conditions changes continuously, but when outline drawings are prepared of many individuals from one clone, and these drawings are compared with similar drawings of individuals from another clone, noteworthy and typical differences become evident between the shapes assumed by the different clones. Characteristically *Amoeba proteus* of the strain T1, reported here, tends to form few and large pseudopodia with smooth outlines. On the other hand *Amoeba discoides* (strain T,) tends to form many pseudopodia, taking on a relatively sheet-like form and having a serrated outline. Thus it is possible, by making many outline drawings of individuals from these two clones and also from the two clones obtained by exchanging nuclei between them, to classify each drawing as typical proteus, typical discoides, more like proteus than discoides, or more like discoides than proteus. The results obtained by examining four clones in this way are shown on Fig. 1. It will be observed that the parent proteus clone has a very high proportion of animals which closely approximate to a monopodal type, but has a certain number of animals which fall into intermediate types or may even transiently resemble the *discoides* type. A similar situation is found for the population of individuals found in a *discoides* clone. The results obtained with these two clones are strikingly different and any two clones can with certainty be classified as *discoides* or *proteus* on this basis. It is equally striking that the heterotransfer clones differ emphatically

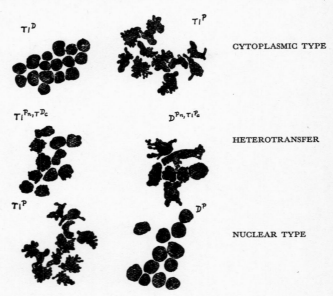

Fig. 2. Silhouettes, taken from above, of amoebae of four clones, obtained by photographing amoebae in paralytic concentrations of clone-specific antiserum obtained from rabbits.

from both the parent stocks. The clone having a *proteus* nucleus in *discoides* cytoplasm rather more resembles the *discoides* clone than the *proteus* clone. Similarly the clone having a *discoides* nucleus in *proteus* cytoplasm displays a rather more striking effect of the cytoplasm than of the nucleus. On the whole, however, the influences of cytoplasm and nucleus on the shape assumed when moving are fairly evenly balanced.

The effect of antisera upon amoebae is also of assistance in studying inheritance. Fig. 2 shows silhouettes obtained by photographing amoebae of various clones which had been exposed to clone-specific antisera. The effect of exposure to a moderate dilution of antiserum is to paralyze the amoebae. When paralyzed they take up shapes which are characteristic of a type or strain. Thus *Amoeba discoides* of Sister Monica Taylor's strain approximates to a sphere, whereas the Tl strain of her *Amoeba proteus* remains with fully extended pseudopodia. As will be seen from the figure, heterotransfers are intermediate, having shapes which lie roughly half-way between those of nuclear and cytoplasmic types.

We have also studied the capacity of various clones of amoebae to become resistant to strain-specific antisera. Our studies of this property are far from complete, but to date it has appeared that only those clones that carry nuclei derived from our Tl strain of Sister Monica Taylor's *Amoeba proteus* are capable of developing a striking resistance to the action of antiserum. The obvious explanation is that it is the nucleus only that is important in determining the capacity to become resistant. This may indeed be the correct explanation. But an alternative explanation is that both cytoplasm and nucleus are involved in the determination of the possibility of becoming resistant, but that so far all the cytoplasms which we have tested have been competent, whereas only the Tl strain of nuclei is competent. The general conclusion which has been drawn from these and other studies made by nuclear transfer is that morphological and physiological characters normally exhibit control both by nucleus and by cytoplasm. It is only when one is investigating the range of macromolecules which can be synthesized by a cell, that all the determinants involved appear to be carried by the nucleus. The cytoplasm and the environment clearly have some importance in determining which of the total range of macromolecules, synthesis of which is rendered possible by the structure of the nucleus, are in fact actually present at a given time. Apart from this, the action of the cytoplasm comes out clearly whenever the characteristic studied is that of an organelle, i.e. of a specific array of macromolecules. In consequence, the hypothesis upon which we are now working is that the cytoplasm controls the way in which macromolecules are organized into functional units. At present genetical theory makes no provision for this essential activity.

Cytoplasmic inheritance and the rate of evolution

If there are separate determinants of inheritance in the cytoplasm, the question at once arises whether the nuclear, or the cytoplasmic determinants, or both, are responsible for the rate of evolution of any particular species. It is far from possible at the present time to provide any definitive answer to this question. In particular it is still uncertain whether cytoplasmic inheritance is of the same importance in sexual reproduction as it is in the case of vegetative reproduction. It is possible that nuclear transfer and other studies will eventually provide the necessary information. At the present time all that can be said is that there are some examples of sexual reproduction in which cytoplasmic inheritance is of outstanding importance, e.g. the painstaking work of Michaelis. But on the other hand there are many instances in which reciprocal crosses fail to reveal any element of cytoplasmic inheritance. This failure may be a reflection of the fact that cytoplasmic inheritance is not important in sexual reproduction. But on the other hand it may also be a reflection of the fact that the cytoplasms of the eggs of the different species between which the crosses were made were, to all

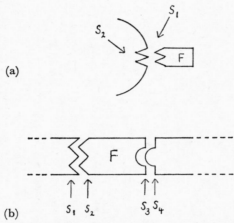

Fig. 3. Diagrams to illustrate the necessity for complementary changes in the surfaces of molecules in the evolution of new structures. (a) The case of variation in a macromolecule attached to an organelle. (b) Case of variation in a macromolecule forming part of a membrane.

intents and purposes, identical from the point of view of inheritance. It is a curious fact that chromatin of one species is only acceptable to egg cells of closely related species. Relatively little attention has been paid to this curious phenomenon. The suggestion is usually made that there is some physico-chemical incompatibility between the chromatin of one species and the egg cytoplasm of another species which is unable to accept that chromatin. This, however, may well be an over-simplification of the facts. The failure of crosses between species that are not closely related may be due to the fact that the plasmagenes of the one species are incompatible with the nuclear genes of the other species. Until experimental evidence is available which can decide this question all other considerations must be to some extent speculative.

It should, however, be noticed that one is just as much justified, so far as present evidence goes, in assuming that the rate of evolution is controlled by the cytoplasm, as in the more commonly accepted assumption that the rate of evolution is controlled by the nuclear genes.

It has been pointed out elsewhere (1953) that there is no substantial evidence that the rate of evolution has ever been controlled by the rate at which new macromolecules can be made available to a species as a result of the rate of evolution (i.e. of mutation) of its nuclear genes. Since the main ascertainable function of nuclear genes is to control the provision of new macromolecules, there must either be some other major activity of nuclear genes or of gene sets which controls the rate of evolution, or the rate of evolution must be controlled by a mechanism not involving the chromosome genes directly. It is also of interest to note that the evolution of higher organisms is mainly dependent not upon the evolution of basic biochemical mechanisms, but upon the evolution of structures. In the formation of new structures we are concerned essentially with the formation of cell organelles, i.e. with those cellular units for which there is already such substantial evidence of cytoplasmic control in vegetative reproduction.

With the aid of Fig. 3 I can make one rather elementary suggestion as to what the relationship may be between nucleus and cytoplasm in the determination of the evolution of new structures. Consider first Fig. 3a; here we have a macromolecule with a function F which is part of a cell organelle. The capacity of the molecule F to form part of the organelle depends upon the complementary character of the two surfaces S1 and S2. So long as the surface S2 remains unchanged, mutation in the F determining gene may proceed with modifications in the function of the compound F. But if any change occurs involving a change in the surface S2, the structure involved will not have the potentiality of being an effective structure unless the surface S1 changes at the same time. Thus for the formation of a new structure, as opposed to a modification of the function of the existing structure, simultaneous mutations affecting both S1 and S2 will be involved. If R is the average rate of mutation of a gene affecting the surface of a macromolecule, then the rate of formation of new structures of this type will be proportional not to R but to R^2.

Now turning to Fig. 3b, where the macromolecule of function F is an integral part of a membrane, we see that the situation may often be much more complicated than this. The structure may be modified so as to insert a new molecule in place of F, only if both S1 and S4 are changed at the same time. This rate of change will be proportional to R^2. But the incorporation of many types of macromolecule into the membrane will require either a change in S1, S2 and S3, i.e. be proportional to R^3, or may even require changes in S1, S2, S3 and S4, i.e. be proportional to R^4. Thus we see that whereas a rate of provision of new types of macromolecule for a cell is proportional to R, the rate of formation of new structures involves a higher power of R. Thus a reasonable working hypothesis is that the rate of evolution of new structures or organelles is the rate determining factor in evolution.

The work described here has been carried out by numerous members of the Department of Zoology at King's College, including Dr. I. J. Lorch, Dr. M. J. Ord, Dr. E. Wilson and Miss S. Hawkins. It has been made possible by grants from the British Empire Cancer Campaign and the Agricultural Research Council, and particularly from the Nuffield Foundation. Some of the apparatus required has been purchased with grants from the Royal Society and the Rockefeller Foundation.

REFERENCES

1. COMMANDON, J. and DE FONBRUNE, P. *Ann. Inst. Pasteur* **60**, 113 (1938).
2. DANIELLI, J. F. *Symp. Soc. Exp. Biol.* **7**, 440 (1953).
3. DANIELLI, J. F. *Proc. Roy. Soc.* B **148**, 321 (1958).
4. DANIELLI, J. F. *Proc. N.Y. Acad. Sci.* (in press) (1959).

LAMPBRUSH CHROMOSOMES

H. G. Callan, L. Lloyd

Department of Natural History, The University, St. Andrews, Fife

Summary

The primary genetic component of a newt lampbrush chromosome is a continuous DNA fibre. This runs out as an axis in each lateral loop, and at each loop base it is compacted to form a chromomere. Aggregated about loop axes are the immediate products of genetic synthesis; these have locus-specific morphologies. An inherent polarization of the primary genetic material is suggested by the regular asymmetry of each lateral loop. Fusion between accumulating gene products leads to the establishment of various intra- and inter-chromosomal relationships; each fusion involves like materials. Lateral loop and centromere morphologies differentiate newt subspecies; these characteristics are conserved in the chromosome complements of first generation hybrids. Lateral loop characteristics also differentiate individuals within a subspecies; these directly reflect allelic differences at certain loci.

Lampbrush chromosomes, like the salivary gland chromosomes of many dipteran larvae, are of exceptional size. Whereas the study of salivary gland chromosomes has been extensive and has contributed significantly to the science of genetics, lampbrush chromosomes have been largely neglected. There is a technical reason why this should be so. Salivary gland chromosomes may be prepared for examination by the now conventional "squash" technique. The st[udy of] lampbrush chromosomes calls for different methods, involvi[ng   ]r of unfixed nuclear contents into saline [8], and their o[bservation by phase] or interference microscopy. Moreover, if entire co[mplements of] lampbrush chromosomes are to be examined a[t one time,] the optical train of the microscope must be inv[erted; the chro]mosomes, falling by gravity in a chamber w[ith a glass-bottomed] coverslip, may be examined from below [12]. [With a phase] contrast microscope, the preparative procedu[re renders study] of these remarkable objects most rewarding.

Lampbrush chromosomes are present in the germinal vesicle nuclei of animal oocytes during the extended prophase of the first meiotic division; in newts, this prophase lasts one or more years. Lampbrush chromosomes have been clearly identified in the oocytes of animals from several vertebrate groups—fish, amphibians, reptiles and birds;

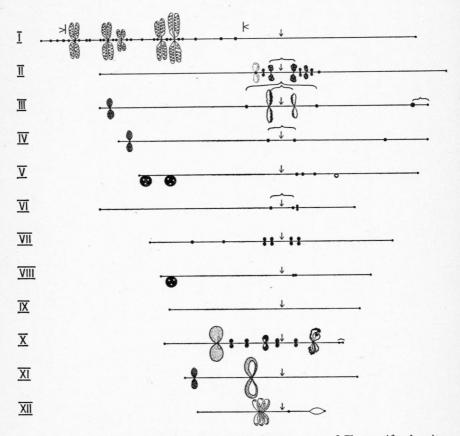

Fig. 1. Working diagram of the lampbrush chromosomes of *T.c. carnifex* showing the relative lengths of the twelve chromosomes and the main "landmarks" used for identification. Centromere positions are marked by vertical arrows. Objects linked by brackets are those which frequently take part in "reflected" fusion. The region of chromosome I demarcated by horizontal arrows is that within which no chiasmata have been observed.

and from at least two invertebrates, the cephalopod mollusc *Sepia* and the isopod crustacean *Anilocra* [2]; those of newts are the largest known. the following account attention will be confined to newt material. opean newts all have a diploid chromosome number of 24, and their oocyte nuclei contain 12 bivalent lampbrush chromosomes. Homologous chromosomes forming a bivalent are joined together at one or more points along their lengths. As will become apparent later, some of these points of association are chiasmata, whereas others are positions where "gene products" fuse.

The axis of each lampbrush chromosome is occupied by a series of small dense chromomeres connected to one another by a thin, extensible and elastic filament [9, 11, 20]. In a fixed preparation these axial chromomeres stain Feulgen-positive [7, 12, 15]. Loops, or other objects of considerable variety of form and texture, project laterally from the great majority of axial chromomeres. These various structures lateral to the axes have morphologies characteristic of specific loci and they are therefore of value for chromosome identification (Fig. 1): in fixed preparations they stain Feulgen-negative. Some of the chromomeres subtend spherical objects, which, if they are small, lie in the axes of chromosomes. In fixed and stained preparations each of these apparently large chromomeres (hereafter termed "axial granules") can be shown to consist of a small crescent of Feulgen-positive material attached to a round mass of Feulgen-negative material. All chromosomes terminate in such axial granules. At a few loci in certain newts, much larger Feulgen-negative spherical objects [1] are similarly attached to crescentic Feulgen-positive chromomeres; these larger objects (hereafter called "spheres") are clearly lateral to the chromosomes' axes.

The majority of lateral loops have a relatively uniform appearance, though they may differ considerably in size from one locus to another. Such loops have a faintly granular texture and show a characteristic asymmetry, each being progressively thinner and denser towards one of the insertions in an axial chromomere, and progressively wider and looser toward the other. In those regions where the texture of a loop is relatively loose, signs of a more refractile loop axis are often apparent (Fig. 2 (1)). In some oocytes (perhaps reflecting enhanced physiological activity) such "typical" lateral loops may carry tiny, very refractile, round granules embedded in their faintly granular and open matrix. These dense granules frequently appear to be arranged in a spiral within the loop matrix, and the more compact parts of the lateral loops are sometimes themselves wound in a spiral. Amongst the more usual types of loops are others of distinctive morphology. Some loops are distinctive in possessing a uniformly compact, stiff matrix, which obscures any signs of a loop axis; others have a stiff foamy texture; others again are remarkably long and thin coiled structures. There is, indeed, very great diversity of lateral loop morphology.

All loops of distinctive appearance can be seen to occur in pairs, or multiples of pairs, at their characteristic sites on the lampbrush chromosomes [1, 15]. Whilst it would not be possible to convince oneself

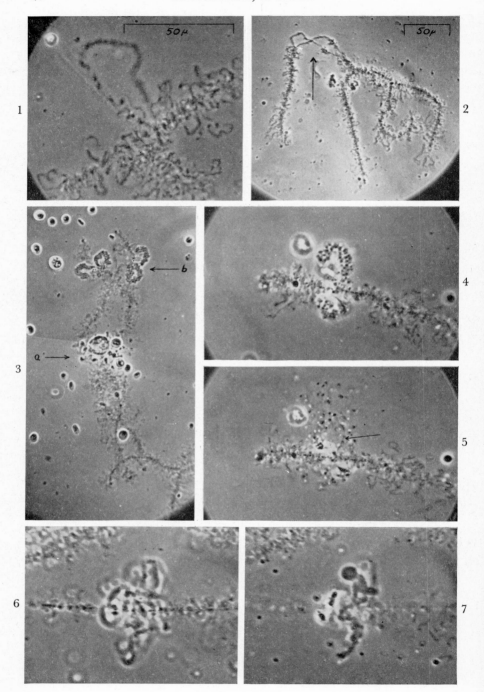

FIGURE 2

by direct observation that the other, less distinctive, lateral loops are also attached in pairs or multiples of pairs to their chromomeres of origin, that such is the case can be shown indirectly. When unfixed lampbrush chromosomes are stretched and broken mechanically, either by accident or design, axial breaks are found to be regularly spanned by a pair of lateral loops ([1,2]: Fig. 2 (2); Fig. 6c). All such axial breaks occur across chromomeres, and between the loop insertions in these chromomeres. The regularity with which double connecting bridges, formed by a pair of straightened-out lateral loops, span axial breaks indicates clearly that all lateral loops are present in pairs, and even in the very frequent situations where many loops take origin in an apparently single chromomere, such multiple loops are in fact arranged sequentially in pairs.

In an unbroken lampbrush chromosome it is generally impossible to determine the "polarity" of a pair of lampbrush loops i.e. how their asymmetry is disposed with respect to the chromosome as a whole. This is, however, immediately possible wherever an axial break has occurred. In the case of several loops of distinctive morphology which can be identified in different oocytes of one animal, and in different individuals of the same species, axial breakages have been observed sufficiently often to allow us to say that the polarity of loop asymmetry is constant. Thus, for example, the giant pair of loops on the longer arm of chromosome XI of *Triturus cristatus carnifex* (Laurenti) always has its thin insertions directed towards the centromere of this chromosome and its thicker insertions directed towards the chromosome end (Fig. 6c). Moreover, in at least one multiple loop of distinctive morphology,

---

Fig. 2. (1) *T.c. carnifex*. Long, but in other respects "typical" lateral loop on heteromorphic arm of chromosome I, with some indication of a loop axis. The sister loop is out of focus.

(2) *T.c. carnifex*. Bivalent XI entire. Axial breakage has occurred across the locus of origin of the giant loops of one homologue, producing the characteristic "double-bridge" marked with an arrow.

(3) *T.c. cristatus*. Bivalent XII entire. The double-axis ends lie below. Half-way up the bivalent lie the giant loops (a), with free products nearby, whilst near the top of the bivalent lie the subterminal giant granular loops (b). Magnification as (2).

(4) *T.c. cristatus*. Subterminal giant granular loops of chromosome XII. Magnification as (1).

(5) *T.c. cristatus*. As (4), but six minutes after addition of ribonuclease at effective concentration 0.1 mgm./cm^3. The loop matrix has dissolved, the granules are dispersing, and a loop axis, marked with an arrow, has been revealed.

(6) *T.c. karelinii*. Large, complex and irregular loops on the heteromorphic arm of chromosome I, which are characteristic of this subspecies: focussed on the axial chromomeres. Magnification as (1).

(7) *T.c. karelinii*. As (6), but focused on the region of the loops from which free products are shed.

namely the giant loop on chromosome XII of *T. c. carnifex* (Fig. 8, (4 and 5), Fig. 9), the polarity of successive loops within the complex is similar and is constant, with the thin insertions directed towards the chromosome end and the thicker insertions directed towards the centromere.

Until recently there have been two diametrically opposed views on the relationship between the axes of lampbrush chromosomes and their lateral structures. Duryee [9, 10] has asserted that the lateral structures are wholly gene products, which are periodically shed from the chromosomes and replenished by further synthesis. Ris [17, 18, 19] on the contrary has maintained that the lateral loops are part of a continuous chromonema, not "lateral secretions of the chromomeres", and indeed in his first paper on the subject Ris altogether denied the existence of a chromomeric axis in lampbrush chromosomes. In their extreme forms neither Duryee's nor Ris' theories fit the facts, yet both have partial validity. Duryee's view is supported by two lines of evidence:

(1) Objects with morphologies similar to many of the structures which are attached laterally to the chromosomes are also found free in the nuclear sap (though objects similar to "typical" loops are not found free unless they have been accidentally detached during preparation).

(2) All the lateral structures disappear, leaving intact chromosome axes, when lampbrush chromosomes are exposed to dilute saline or proteolytic enzymes. Moreover, towards the end of oocyte development the lampbrush chromosomes "lose" their lateral structures, shorten, and become normal diakinetic chromosomes.

Ris' view gains support from the mere fact that most of the lateral structures have the form of loops. With the exception of the spheres, even those lateral structures which are not apparently loops reveal an inherent loop structure at certain stages of oocyte development and during dissolution in saline of low concentration (Fig. 8 (8)). We may dismiss Ris' claim that lampbrush chromosomes lack axial chromomeres, and in fact Ris has recently admitted [19] that there are "heterochromatic regions in which the chromonemata are tightly coiled" in lampbrush chromosome axes. However, if the lateral loops are simply chromonemata we are entitled to wonder how it comes about that they have markedly different morphologies at different chromosomal loci, and why some of the lateral objects show no signs of a fibrillar structure.

A reasonable interpretation of the relationship between lampbrush chromosome axes and their lateral loops lies between these extreme views. There are grounds for supposing that although the main bulk of each lateral loop consists of gene product (loop "matrix"), each loop has an axis constructed from chromomeric material, and this axis forms part of a continuous DNA fibre which runs throughout the length of a lampbrush chromosome. This interpretation was at first based on indirect evidence [1].

Although the weakest parts of lampbrush chromosomes would appear to be the thin filaments between chromomeres, stretching produces axial breaks within chromomeres which always cause pairs of loop insertions to become disengaged. This suggests that each chromomere consists of at least two separable parts *along the chromosome axis*. Loops that at first sight appear to be composed entirely of granules can form persistent "double bridges" spanning axial breaks. This suggests that even granular loops have a filamentous component. Broadly, considered throughout oocyte development, lateral loop size and chromomere size are inversely correlated, most loops being longest and chromomeres smallest in oocytes of about half their mature diameter. Moreover, the giant loops spring from exceedingly tiny chromomeres, within which axial breakage occurs disproportionately often. Thus, in spite of the evident chemical difference between Feulgen-positive chromomeres and Feulgen-negative lateral loops, a constituent which contributes in minor degree only to the bulk of the loops is probably derived from chromomere substance.

The direct evidence supporting this interpretation may be stated as follows:

(1) When lampbrush chromosomes are exposed to peptic digestion the lateral structures disappear. However, if the loops are first anchored to a film by fixation and then subsequently treated with enzyme, pepsin-resistant loop axes of approximately the same thickness as the interchromomeric filament can be demonstrated with the electron microscope [14].

(2) Ribonuclease also promotes the solution of lateral loop matrices but its action, at physiological pH, is much less drastic than that of pepsin in the presence of strong acid. Near the end of the long arm of chromosome XII of *T. c. cristatus* (Laurenti) is a characteristic pair of giant loops having matrices of loose, open texture in which are embedded some 50–100 granules each $1-2\,\mu$ in diameter. The thin insertion of this loop is a spirally wound dense filament, but in the remaining part of the loop no axis is normally visible (Fig. 2 (3 and 4)). If this loop is kept under observation whilst ribonuclease digestion proceeds the matrix dissolves, the granules are set free and an inner dense loop axis is revealed (Fig. 2 (5)). This loop does not respond to ribonuclease digestion in an exceptional manner: it is, however, a particularly favourable test object on account of its size and texture.

(3) Deoxyribonuclease acts on lampbrush chromosomes in a dramatically different manner. This enzyme causes interchromomeric fila-

ments to break, *and it also breaks up the lateral loops into tiny fragments* [4]. If the progress of deoxyribonuclease digestion is watched down the microscope, it is apparent that this enzyme does not initially disturb the fine texture of the lateral loops: the loops merely become chopped up into smaller and smaller portions. This is presumably due to fragmentation of their axes.

The evidence in favour of the lateral loops possessing DNA axes, and

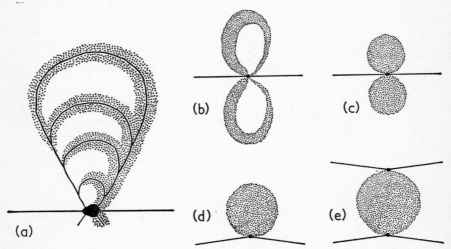

Fig. 3. (a) Diagram illustrating a theory advanced to explain the characteristic asymmetry of lateral loops. Four successive stages of loop development are drawn and the loop axis is assumed to extend from one side of the parent chromomere only. The most recently extended parts of loop axis, shown on the left of the diagram, are bare of loop matrix, whereas the "older" parts are coated with matrix.
  (b), (c), (d) and (e) are diagrams illustrating matrix fusion:
  (b) sister lateral structures separate and loop form conserved.
  (c) sister lateral structures separate but loop form obliterated by fusion.
  (d) sister lateral structures fused together.
  (e) sister and inter-homologue fusion.

for the interchromomeric fibrils to be similarly constituted, is thus conclusive. In consequence we must visualize each chromosome as a continuous DNA fibril several centimetres in length [14]—perhaps not only in oocyte nuclei but also in somatic "resting" nuclei of volume a few thousand cubic microns!

We consider the DNA constituents of lampbrush chromosomes—interchromomeric fibrils, chromomeres and lateral loop axes—to be the "persistent" genetic material, and the loop matrices, wholly or in part, to be gene products. Loop matrix accumulates on a loop axis as this axis extends from a chromomere; periodically, portions of matrix are shed and come to lie as free bodies in the nuclear sap. When first shed, the texture and form of these free objects may be sufficiently distinctive to be identified with particular loci on the chromosomes. Later these objects undergo metamorphoses, with which we are not concerned in the present paper.

Two possible explanations for the characteristic asymmetry of lateral loops come to mind. Each loop axis may contain a sequence of genetic information along which synthesis proceeds in stages, the finished product accumulating at the thick end of the loop. Alternatively, each loop axis may comprise a series of repetitions of the same genetic information: if a chromomere be so polarized that loop extension occurs progressively from one point of insertion only [1, 13], there will be "young" and "old" parts of a lateral loop, with least accumulation of loop matrix in the youngest region (Fig. 3a). For several reasons we favour the latter explanation, in particular because the products of loops of a wide range of morphologies are not shed from restricted parts of loops close beside the chromosome axis. Thus in the case of certain easily recognized loci, such as the large irregular "heteromorphic" loops on chromosome I of *T. c. karelinii* (Strauch) (Fig. 2 (6 and 7)) the product tends to be shed from a region about half way along the loop—i.e. from the part of the loop farthest removed from the chromosome axis, with the consequence that although such loops have the usual thin and thick insertions in a chromomere, the thick insertion is not necessarily the thickest part of the entire loop. This suggests that all parts of a loop are carrying out the same synthesis, but that different parts have been engaged in synthesis for differing periods of time.

Mention has already been made of the fact that whereas most of the structures lateral to the axes of lampbrush chromosomes take the form of loops, objects of other morphologies are also present at certain loci. With the exception of the "spheres" these other objects arise by fusion of loop matrix, a process which can lead to several different situations (Fig. 3b–e). The matrices of a pair of sister loops may separately fuse to form two irregular masses attached on either side of an axial chromomere. Alternatively the matrices of a pair of sister loops may fuse together, in which case the locus is occupied by a single irregular (though in some instances more or less spherical) mass lateral to the chromosome axis. Frequently such structures betray their origin from a pair of lateral loops; the twin loop insertions project as a pair of "tails" from the mass and bend sharply back to their parent chromomere. Another situation is encountered when the matrices of multiple loops fuse together, as is notably the case with the giant loops near the middle of chromosome XII of *T. c. carnifex* (Fig. 8 (5)) and *T. c. karelinii*: full analysis of such a complex structure is often not possible.

Yet another situation is provided by matrix fusion between corresponding loops on homologous chromosomes, (Fig. 4 (1); Fig. 10d),

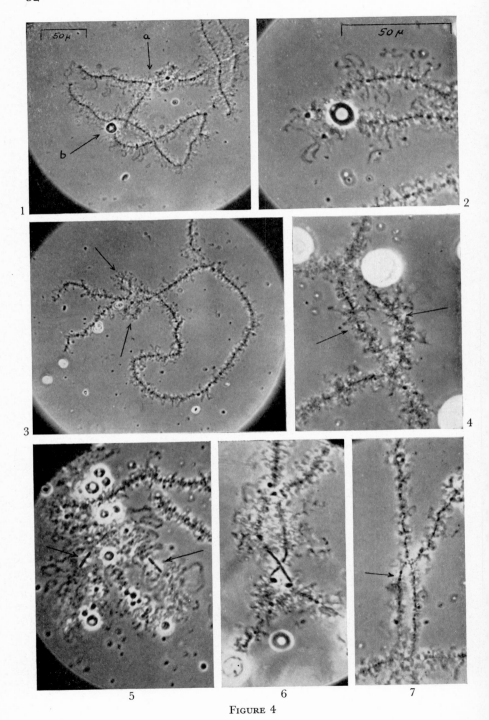

Figure 4

as frequently occurs with the "lumpy" loops around the centromere of chromosome II of *T. c. carnifex* and of *T. c. karelinii*, and with the giant loops near the middle of chromosome XII of *T. marmoratus* (Latreille) [1]. This is a form of homologue union to be distinguished from chiasmate associations involving lampbrush chromosomes' axes, though it may possibly have genetical implications.

We interpret axial granules and spheres as exceptional gene products only to the extent that they are produced beside compact chromomeres rather than in association with extended lateral loops. As with loop matrices, fusion between axial granules and between spheres may occur, and it gives rise to several characteristic situations. Thus fusion between homologous terminal or other axial granules is frequent; so is fusion between homologous spheres on chromosomes V (Fig. 5b-d) or VIII (Fig. 4 (2)) of *T. c. carnifex*. However, non-homologous fusion also occurs; for example, a terminal axial granule of chromosome X fused to one of the many interstitial granules present on the long arm of chromosome I of *T. c. carnifex* is shown in Fig. 4 (1).

Another type of non-homologous association involving axial granules within one and the same chromosome leads to the production of looped chromosome axes [12], and for descriptive purposes we term this "reflected fusion". Chromosome VI of *T. c. carnifex* has two axial granules symmetrically placed a short distance apart on either side of its centromere; fusion between these granules occurs so frequently as to

---

Fig. 4. (1) *T.c. carnifex*. Bivalent X entire, with a terminal granule (a) non-homologously fused to an interstitial axial granule on the heteromorphic arm of chromosome I. Just to the right of the arrowhead lies one of the multiple coiled loops characteristic of the heteromorphic arm of chromosome I of *T.c. carnifex*. The giant loop locus of chromosome X, present on both homologues in this individual and showing interhomologue fusion in this particular nucleus, is marked (b).

(2) *T.c. carnifex*. Interhomologue fusion at the "sphere" loci of bivalent VIII. The terminal granules are also conspicuous in this photograph.

(3) *T.c. carnifex*. Bivalent VI entire, with both homologues showing pericentric reflected fusion of axial granules. The centromere loci are indicated by arrows. Magnification as (1).

(4) *T.c. carnifex*. Part of bivalent I including the centromeres, which are marked by arrows. The large white objects are free "nucleoli". Magnification as (2).

(5) *T.c. karelinii*. Part of bivalent VII including the centromeres, which are marked by arrows. The highly refractile round objects are "lumpy" loops. Magnification as (2).

(6) *T.c. karelinii*. Part of bivalent XII showing fusion between the centromeres. The round object at bottom left lies at the giant multiple loop locus. Magnification as (2).

(7) F_1 ♀ hybrid *carnifex* ♀ × *karelinii* ♂. Part of bivalent X including the centromeres. The *karelinii*-derived centromere is marked with an arrow: the *carnifex*-derived centromere lies at the homologous locus on the partner chromosome just below the chiasma, but is inconspicuous. Magnification as (2).

be a useful diagnostic feature of this chromosome (Fig. 4 (3)), the centromere lying half-way along the looped portion of the axis. Reflected fusion between axial granules further apart but also symmetrically disposed on either side of the centromere of chromosome III of *T. c. carnifex* is again an almost invariable character; less frequent pericentric reflected fusion occurs in chromosome IV of this subspecies. In *T. c. karelinii*, chromosome II, pericentric reflected fusion of axial granules situated just beyond the zone of "lumpy" loops appears to be invariable, though oddly enough the very similar chromosome II of

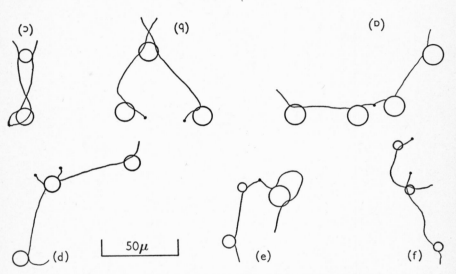

Fig. 5. Camera-lucida drawings of the "sphere"-bearing end of bivalent V of *T.c. carnifex* from six different oocytes, showing chromosome axes and spheres only.
(a) Fusion between terminal axial granules.
(b) Fusion between proximal spheres.
(c) Fusion between terminal axial granules, distal and proximal spheres.
(d) Fusion between distal spheres.
(e) Fusion between terminal axial granules and reflected fusion between the distal and proximal spheres of one chromosome.
(f) Non-homologous fusion between distal sphere of one chromosome and proximal sphere of its partner.

T. c. carnifex never shows fusion between these granules. Reflected fusion of terminal with subterminal axial granules is also encountered, notably within the shorter arms of chromosomes III and X of *T. c. carnifex*. This leads to the termination of a chromosome in an axial loop (Fig. 10d): a situation not to be confused with the "double axis end" of chromosome XII of *T. marmoratus* and of all subspecies of *T. cristatus* [1].

The two spheres near the end of one arm of chromosome V of *T. c. carnifex* occasionally show reflected fusion, and indeed the four spheres on the two chromosomes forming bivalent V can fuse in all conceivable combinations (Fig. 5). Furthermore, very occasionally, fusion takes place between a sphere on chromosome V and another on chromosome VIII, and this raises the problem as to whether any of these fusions are between genuinely non-homologous materials.

For fusion of gene products (loop matrices, axial granules or spheres) to occur it is clear that at least three conditions require to be fulfilled:

(1) The gene products must accumulate where they are being synthesized, and not diffuse away through the nuclear sap from their site of origin as fast as they are formed. For many loci this first condition may never be met.

(2) The materials synthesized at two sites must accumulate in close proximity to one another within the nucleus. Cross-fusion of the products of partner chromosomes will evidently be favoured, since at zygotene these chromosomes were intimately paired. The high frequency of pericentric reflected fusion in certain chromosomes is perhaps a result of the maintenance of the anaphase arrangement of the chromosomes from the preceding mitotic division within the young meiotic nucleus: with the centromeres leading towards the spindle poles, the two arms of a chromosome necessarily come to lie close to one another.

(3) The texture of the gene products must be such that they flow together: their surface architecture may also be of critical importance. Some loop matrices never fuse; this is true of the granular loops near the end of the longer arm of chromosome XII of *T. c. cristatus*, where even the granules within a single loop remain separate. Some loop matrices regularly fuse in young oocytes, but are generally separate in older oocytes: this is the case for the giant loops on chromosome XI of *T. c. carnifex*, which show a change of texture with increasing oocyte age (Fig. 6a-d).

Taken overall, our general impression is that fusion of gene products is material-specific, and that whenever fusion occurs between loci that are not genetically homologous—in the sense that the loci involved do not occupy corresponding places on partner chromosomes—the fusing gene products are nevertheless chemically similar if not indeed identical. The degree of development, texture and vacuolation of the spheres in *T. c. carnifex* varies according to oocyte age and other factors as yet little understood, but in a particular oocyte, whatever characterizes the spheres on bivalent V characterizes also those on bivalent VIII: it is probable that all the loci responsible for sphere production are physically and chemically identical. Spheres fuse with other spheres, but not with gene products having other morphologies. Likewise all

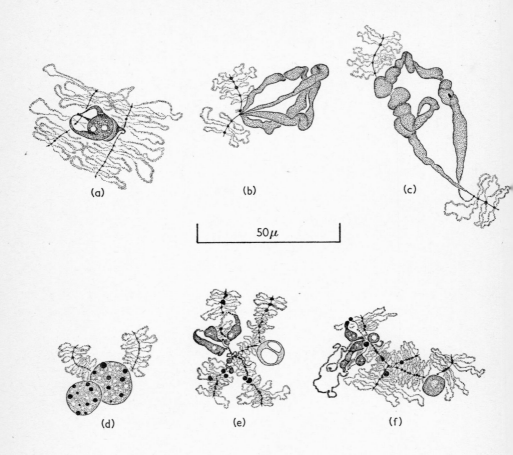

Fig. 6. Camera-lucida drawings of the giant loops and neighbouring regions of chromosome XI of *T.c. carnifex*.

(a) Fusion between sister and homologous loop matrices, from an oocyte of 0.9 mm. diameter. This condition is frequently encountered in small oocytes.

(b) Sister loops separate, from an oocyte of 1.1 mm. diameter. This is the typical condition for oocytes midway through their period of growth.

(c) Axial breakage at the giant loops locus, from an oocyte of 1.2 mm. diameter. The chromosome axis at bottom right runs towards the centromere.

(d) Fusion between sister loop matrices, with a free body about to be shed, from an oocyte of 1.3 mm. diameter. This condition, in which the loop form is entirely obliterated, is typical for oocytes nearing maturity.

(e) and (f) show the heteromorphic condition at the giant loops loci typical of ♀ R, with sister loops unfused on one chromosome, fused together on its partner. (e) is from an oocyte of 1.2 mm. diameter, (f) from an oocyte of 0.8 mm. diameter.

terminal axial granules may reasonably be supposed to be identical, and at least very similar to those interstitial axial granules with which they occasionally fuse.

The axial granules near the centromeres of chromosomes III, IV and VI of *T. c. carnifex* do not precisely resemble other axial granules and in particular they have less regular outlines. Their reflected fusion could be due to genetic identity. Several of the chromosomes of

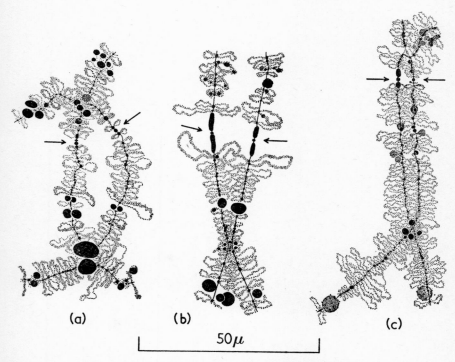

Fig. 7. Camera-lucida drawings of the centromeres and neighbouring regions of bivalents X. The centromeres are marked by arrows: parts of the shorter arms of chromosomes X lie above the centromeres in all three drawings.
  (a) *T.c. carnifex*, from an oocyte of 1.2 mm. diameter.
  (b) *T.c. karelinii*, from an oocyte of 1.1 mm. diameter.
  (c) *carnifex* ♀ × *karelinii* ♂, from an oocyte of 0.9 mm. diameter.

T. c. carnifex, notably II, III, IV, VI and VII have, at least to a first approximation, symmetrical dispositions of recognition characters about their centromeres. The symmetry could be accounted for if these chromosomes were originally isochromosomes resulting from centric misdivision [6, 16].

When Gall characterized the eleven lampbrush chromosomes of the

American newt *Triturus viridescens* [12] he made use of centromere locations for chromosome identification. In this species, and also in *Amblystoma tigrinum*, the centromeres (= kinetochores) are conspicuous, being long (ca. $10\,\mu$) dense axial structures devoid of lateral loops. The centromeres of *T. c. carnifex*, on the contrary, are inconspicuous objects, and their identification proved exceedingly laborious: they are tiny Feulgen-positive spheres $1-2\,\mu$ in diameter, lying in the chromosome axes or slightly to one side, and lacking lateral loops (Fig. 4 (4); Fig. 7a). In the axes of lampbrush chromosomes of *T. c. carnifex* having maximally developed lateral loops there are admittedly very few chromomeres lacking lateral structures of one kind or another (though axial granules superficially present this appearance). However, amongst the several hundred chromomeres of a single chromosome it is by no means easy to pick out the one which genuinely and regularly lacks lateral loops, and in any case when the study began we had no guarantee that absence of lateral loops would be a centromere characteristic in *T. c. carnifex*.

We do not here propose to describe in detail how the centromeres of *T. c. carnifex* were first found, but a study of the distribution of chiasmata along the lengths of the *carnifex* bivalents provided a most helpful clue. It so happens that in *T. c. carnifex* female meiosis the chiasmata tend to be clustered in regions close to the centromere, so that a histogram of chiasma frequency plotted against length for fifty or more homologous chromosomes shows, somewhere along the chromosome, two humps beside one another. The position of the trough between the humps marks the centromere. In carrying out this study we had stumbled on another characteristic of *carnifex* centromeres: axial fusion *never* occurs at the centromeres of this subspecies, a somewhat surprising observation since Gall [12] had clearly demonstrated occasional "kinetochore fusion" in *T. viridescens*.

It was only after several months occupied in finding the centromeres of *T. c. carnifex* that we started to study the lampbrush chromosomes of *T. c. karelinii*, which to our considerable chagrin proved to have centromeres fully as conspicuous as those of *T. viridescens*. However the chromosome complements of *T. c. carnifex* and *T. c. karelinii* have an overall similarity, and there was at least some gratification in finding that the *karelinii* centromeres lie in chromosome regions homologous to those occupied by the assumed centromeres of *carnifex*.

The centromeres of *T. c. karelinii* lampbrush chromosomes sometimes have the appearance of uniform, Feulgen-positive, dense axial cylinders, lacking loops and between 10 and $20\,\mu$ long; more frequently each consists of a small loop-free granule separated by constrictions from two $5-10\,\mu$ long dense loop-free axial cylinders lying at either side (Fig. 4(5); Fig. 7b). In *T. viridescens* Gall had already described some centromeres as carrying a granule imbedded in, or associated laterally with, a loop-free axial cylinder, and this form also occurs in *T. c. karelinii*. It was surprising to find the centromeres of two newt subspecies, *T. c. carnifex* and *T. c. karelinii*, so different in appearance, but still more astonishing to find axial fusion between homologous *karelinii* centromeres to be of frequent occurrence (Fig. 4(6)): absolute non-occurrence of axial fusion had been the very criterion by which we first established the positions of *carnifex* centromeres!

The centromeres of the other two subspecies *T. c. cristatus* and *T. c. danubialis* (Wolterstorff) resemble those of *T. c. carnifex* but are smaller in size. In two first generation subspecies hybrids which we have examined, *carnifex* ♀ × *karelinii* ♂ and *danubialis* ♀ × *karelinii* ♂, the parental centromere characteristics are sufficienty conserved to permit ready distinction between homologues (Fig. 4(7); Fig. 7c). Nevertheless the *karelinii* centromeres in a hybrid are less refractile than in a purebred *karelinii* and the *danubialis* or *carnifex* centromeres decidedly more conspicuous than normal. This raises a problem of some interest.

The centromeres—and in the case of *T. c. karelinii* the associated loop free axial cylinders—are Feulgen-positive, just as are chromomeres bearing lateral structures. In *T. c. karelinii* not only are the centromeres conspicuous objects; most of the axial chromomeres are considerably larger than those of the other subspecies when comparisons are made between chromosomes having lateral loops developed to the same degree. This distinction is also evident in fixed Feulgen-stained preparations: in *T. c. karelinii* the Feulgen-positive axial chromomeres can be made out with ease throughout the length of each chromosome, whereas in the other subspecies—particularly *T. c. cristatus* and *T. c. danubialis*, only occasional chromomeres are visible. This suggests that there may be significant differences in the amounts of DNA per oocyte nucleus in the four subspecies, and if the oocyte nuclei are not exceptional one would expect to find corresponding differences between somatic cell nuclei also; this question is now being examined. The distinction between the subspecies is not preserved in F_1 hybrids: thus in Feulgen preparations of the hybrid *danubialis* ♀ × *karelinii* ♂ no general difference is observable between *danubialis*-derived and *karelinii*-derived chromomeres, but the former are decidedly more conspicuous than they are in pure-bred *danubialis*. The study is not yet sufficiently far advanced to warrant further comment at this stage.

The lateral components of the lampbrush chromosomes of the four subspecies of *T. cristatus* have overall similarity, yet features distinctive of each subspecies. We know from studies on male meiosis that the subspecies differ by translocations [5]; multivalent chromosome associations are present also in oocyte nuclei of subspecies hybrids. However, quite apart from distinctive "structural" arrangements of the genetic material, certain types of lateral loops are restricted to particular subspecies. Thus, for example, the giant granular loops near the end of the long arm of chromosome XII of *T. c. cristatus* (Fig. 2 (3 and 4)) are not present on chromosome XII nor any other chromosome within the complements of the other three subspecies.

In all oocytes of all females of all four subspecies chromosome I is peculiar in that the partner chromosomes are never of identical morphology. There is a region of the long arm of this chromosome extending from close to the centromere nearly to the end of the arm, in which chiasmate association has not been observed (Fig. 1). Within this region "heteromorphic" loops are present on one chromosome which are not matched by similar loops on the partner, a situation which suggests that female newts are heterogametic, bivalent I being the sex-determining pair. The most conspicuous heteromorphic loops are of different morphologies in different subspecies, those of *T. c. cristatus* and *T. c. carnifex* being multiple-coiled structures of relatively uniform width (Fig. 4 (1)), those of *T. c. danubialis* being "typical" but exceptionally long, whereas those of *T. c. karelinii* are highly irregular in shape (Fig. 2 (6 and 7)). In the F_1 hybrids *carnifex* ♀ × *karelinii* ♂ and *danubialis* ♀ × *karelinii* ♂ which we have examined, the heteromorphic loops on chromosome I precisely resemble those of pure-bred *T. c. karelinii* ♀♀. If this should be a rule without exception for hybrids whose male parent was *T. c. karelinii*, we may assume that the chromosome bearing the conspicuous heteromorphic loops is represented twice in the male complement.

The faithful conservation of loop morphology from parent to hybrid offspring, which holds for several other distinctive loci, shows that each genetic locus precisely controls the form assumed by its immediate product in oocyte nuclei. The following observations further substantiate this statement.

Fig. 8. All figures of *T.c. carnifex*, and at same magnification

(1) ♀ R. Bivalent XI entire, with sister giant loops separate (a) on one homologue, fused together (b) on the partner chromosome.

(2) As figure 1 but from another nucleus, with an unidentified chromosome lying below bivalent XI.

(3) ♀ E. Bivalent XII entire, lacking giant multiple loops.

(4) ♀ M. Bivalent XII entire with giant multiple loops, marked with arrow, on one homologue only.

(5) ♀ J. Bivalent XII entire with giant multiple loops, marked with arrow, on one homologue only.

(6) ♀ E. Bivalent X entire with giant loops, marked with arrow, on one homologue only.

(7) As (6) but from a nearly mature oocyte. Most of the lateral loops are present as vestiges only, but the giant loops remain conspicuous.

(8) As (6) but from a smaller oocyte and isolated in saline of lower concentration.

LAMPBRUSH CHROMOSOMES

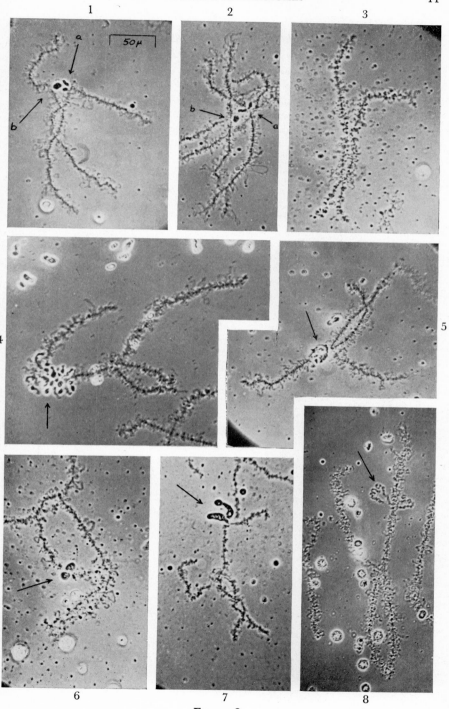

FIGURE 8

The lampbrush chromosomes of no two individuals of *T. c. carnifex* are exactly alike [3]. Some individual-specific characters are subtle. In most specimens of *T. c. carnifex* the matrices of the giant loops on chromosome XI from half-grown oocytes do not fuse sufficiently to obliterate the fundamental loop form, nor are sister loops fused together. However, in one individual, ♀ R, homologous giant loops on bivalent XI are not identical: in every oocyte examined the sister loops were separate on one chromosome but fused together on its partner (Fig. 8 (1 and 2); Fig. 6e and f). Other individual-specific features are much

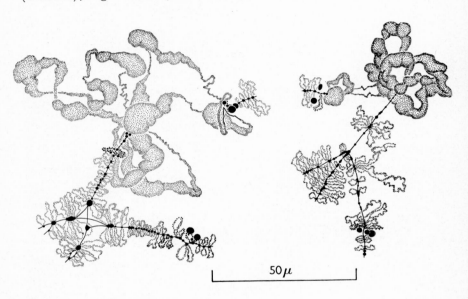

Fig. 9. Camera-lucida drawings of part of bivalent XII from two nuclei of an individual (♀ M) of *T.c. carnifex* heterozygous for giant multiple loops. In both examples a chiasma lies between the giant multiple loops and the centromeres, a region of chromosome XII where the axis is sometimes clearly double. In both examples axial breakage has occurred within the multiple loop complex. Notice that small "typical" loops occupy the zone homologous to the giant multiple loops locus.

more obvious. The multiple giant loops on chromosome XII of *T. c. carnifex* may be absent altogether from every oocyte of a given animal (Fig. 8 (3)), in another individual regularly present on one chromosome but absent from its partner (Fig. 8 (4 and 5); Fig. 9). It happens that we have so far not found a single specimen of *T. c. carnifex* homozygous for the "presence" of these giant loops, though all individuals of *T. c. karelinii* and of *T. c. cristatus* have this constitution. The giant loops on the longer arm of chromosome X of *T. c. carnifex* are also absent in

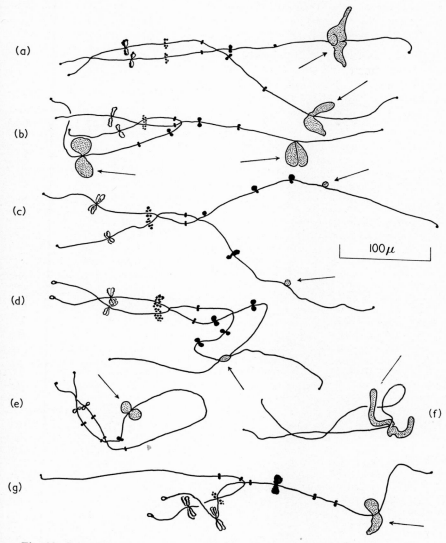

Fig. 10. Camera-lucida drawings of bivalent X from three females of *T.c. carnifex*, showing chromosome axes and main "landmarks" only. Lateral objects at the giant loops loci are marked with arrows.

(a) and (b) from ♀ G, homozygous for the presence of giant loops and scored +/+.

(c) and (d) from ♀ O, homozygous for inconspicuous loops at the giant loops loci and scored —/—. In (d) matrix fusion between homologues has occurred.

(e), (f) and (g) from ♀ E, heterozygous at the giant loops loci and scored +/—.

The heteromorphic development of other landmarks shown in (c), (d), (e) and (g) is characteristic of ♀♀ O and E. The lack of all but one landmark in (f) is due to oocyte maturity.

some animals (Fig. 10 c and d), other animals being heterozygous for this character (Fig. 8 (6-8); Fig. 10, e-g) and yet others homozygous for the presence of giant loops (Fig. 10 a and b). It is not strictly correct to speak of the "presence" or "absence" of giant loops on chromosomes X and XII since these are absolute terms: in a chromosome lacking giant loops the locus is not missing—the loops at the locus in question are merely inconspicuous. The terminology is adopted for the sake of simplicity. Scored as "+" or "−", the constitution of thirteen specimens as regards giant loops on chromosomes X and XII are set out in Table 1, together with figures of expectation based on the assumption that these characters are inherited in Mendelian fashion and assort freely. The tallies show good agreement between observation and expectation.

The above examples of individual-specific characters are only a few of the many that we have so far noticed. Even centromere characteristics may be specific to individuals: thus in one specimen of *T. c. karelinii* homologous centromeres in bivalent VI are regularly of different morphologies, being respectively a long, unconstricted, dense axial cylinder with a laterally attached sphere on one chromosome, and

TABLE 1
Triturus cristatus carnifex

Individual	Chromosome X	Chromosome XII
C	−/−	−/−
D	−/−	−/−
E	+/−	−/−
F	+/−	−/−
G	+/+	−/−
J	−/−	+/−
L	+/−	−/−
M	−/−	+/−
N	+/−	−/−
O	−/−	−/−
P	+/+	−/−
Q	+/−	+/−
R	+/+	+/−

	−/−	+/−	+/+	−/−	+/−	+/+
Observed	5	5	3	9	4	0
Expected	4.3	6.4	2.3	9.3	3.4	0.3

a much smaller axial cylinder with a conspicuous constriction on the partner. Moreover, individual-specific characters such as those described above, heterozygosities being most easily noticed, are persistent in time. If a newt is characterized on the chromosomes from a sample of oocytes removed on one occasion, subsequent samples taken many months later from the same animal conform in their entirety to the earlier characterization. It is thus evident that we are able to recognize allelic differences at particular genetic loci by the form assumed by the immediate products of gene-controlled synthesis, and since we are able to breed from animals so characterized it should not be difficult to establish the modes of their inheritance.

The extent to which correlation exists between cytologically observed genetic differences between newts and characters of newt phenotypes is a much more formidable problem and on this question we have, as yet, no observations to record. Nor do we know anything specific about the chemical natures of the gene products that can be directly observed along the lengths of lampbrush chromosomes. Until we can say that such and such a material, chemically and physically characterized, is manufactured at such and such a locus on a chromosome, there will remain a gap between what we know genetically concerning a unit of inheritance, and what we know, working back from the phenotype, about its role in physiological and developmental processes. In our opinion an exploitation of the remarkable possibilities afforded by lampbrush chromosomes has good prospect of closing this gap, and for this reason if for no other, lampbrush chromosomes deserve intensive study in the coming years.

We are greatly indebted to Dr. H. Spurway, Professor G. Montalenti, Dr. P. Dohrn, Professor F. Mainx and Dr. A. Sengün for supplying the newts used in this investigation; to the late Mr. D. G. Brown for his care in maintaining these newts and their offspring in good condition in the laboratory; to Mr. H. C. Macgregor for the photographs forming Fig. 2 (4 and 5); to Cooke, Troughton and Simms, Ltd., for the loan of Xenon flash equipment, and to Dr. G. A. Horridge for designing and constructing a similar piece of apparatus for our later use.

References

1. CALLAN, H. G. In "Symposium on Fine Structure of Cells." I.U.B.S. publication, Noordhoff, Groningen, ser. B, **21**, 89. (1955).
2. CALLAN, H. G. *Pubbl. Staz. Zool. Napoli* **29**, 329 (1957).
3. CALLAN, H. G. and LLOYD, L. *Nature, Lond.* **178**, 355 (1956).
4. CALLAN, H. G. and MACGREGOR, H. C. *Nature, Lond.* **181**, 1479 (1958).
5. CALLAN, H. G. and SPURWAY, H. *J. Genet.* **50**, 235 (1951).
6. DARLINGTON, C. D. *J. Genet.* **37**, 341 (1939).
7. DODSON, E. O. *Univ. Calif. Publ. Zool.* **53**, 281 (1948).
8. DURYEE, W. R. *Arch. Exp. Zellforsch.* **19**, 171 (1937).

9. DURYEE, W. R. In "University of Pennsylvania Bicentennial Conference on Cytology, Genetics and Evolution." Philadelphia, 129 (1941).
10. DURYEE, W. R. *Ann. N.Y. Acad. Sci.* **50**, 920 (1950).
11. GALL, J. G. *Exp. Cell Res.* Suppl. **2**, 95 (1952).
12. GALL, J. G. *J. Morph.* **94**, 283 (1954).
13. GALL, J. G. *Symp. Soc. Exp. Biol.* **9**, 358 (1955).
14. GALL, J. G. *Brookhaven Symp. Biol.* **8**, 17 (1956).
15. GUYÉNOT, E. and DANON, M. *Rev. Suisse Zool.* **60**, 1 (1953).
16. RHOADES, M. M. *Genetics* **23**, 163 (1938).
17. RIS, H. *Biol. Bull.* **89**, 242 (1945).
18. RIS, H. *J. Biophys. Biochem. Cytol.* **2**, 385 (1956).
19. RIS, H. in "A Symposium on the Chemical Basis of Heredity." Johns Hopkins, Baltimore, 23 (1957).
20. TOMLIN, S. G. and CALLAN, H. G. *Quart. J. Micr. Sci.* **92**, 221 (1951).

THE MORPHOLOGY OF DEVELOPING SYSTEMS AT THE ULTRAMICROSCOPICAL LEVEL

C. H. WADDINGTON

Institute of Animal Genetics, Edinburgh

THE discoveries of genetics have provided a very firm basis for the belief that the ultimate determinants of cellular character are to be found in the nuclear genes. There are, however, still very many gaps in our understanding of how the genes operate. It is clear that the development by a cell of a particular character involves a two-way traffic, from the genes to the rest of the cell and from the rest of the cell to the genes.

We already have some sort of theory as to how the genes transmit information to the remainder of the cell. According to this, the genetic material consists essentially of the sequences of nucleotides composing DNA. These nucleotide sequences are supposed to determine the sequences of amino acids which become coupled together to form the proteins out of which the cellular cytoplasm is constructed. This theory, although intellectually very neat and appealing, is obviously no more than a skeleton. For instance, we know that RNA comes into the story somewhere, but exactly where is still rather obscure. Moreover,—and this is the point that I wish to stress in the present context—we know that the genes are operating in a system which has a complex morphology of membranes, granules and other structures. Any complete theory of genic action must certainly give some indication of how these entities are involved.

Our knowledge of the other direction of information-flow, from the remainder of the cell to the genes, is even less complete. In fact, we can hardly be said to have the beginning of an understanding of what types of process are involved in the decision that in a certain group of cells one particular constellation of gene-controlled actions will proceed and lead the differentiation along one path, while in a different group of cells some other set of activities will lead to a different end result. The study of embryonic induction has demonstrated clearly enough that influences which impinge in the first place on the cytoplasm may suffice to determine which of the alternative pathways will be followed. But we do not know on which elements in the cytoplasm these influences initially operate, nor do we understand the steps by which their action is finally transmitted to the genes. Here again, however, we must expect that any full theory will have to take account of the structural complexity of the cytoplasm.

The development of electron microscopy has revealed a whole world of morphological structure within the cell which was previously beyond the limits of our investigation. An understanding of how this realm of structures enters into the life of the cell will, of course, require something more than a mere inspection of its morphology; biochemical and other experimental techniques will be essential for the investigation. However, there is much to be said for the somewhat oldfashioned view that the first step towards understanding a system is to look at it, and the electron microscope has given us a method of looking at cells which will be neglected at our peril.

In the few years since these techniques have become generally available, rather little has been done to apply them to the study of developing systems in which we have at the same time some knowledge of the physiological processes by which the development is being guided. Some first steps in this field have, however, been taken for instance, by Lehmann on *Tubifex* eggs (1956) and Yamada (1957), Bellairs (1958), Selman and Waddington (1958) on amphibia and chicken embryos. Much of this work has the character of a general survey aimed at discovering what the situation is, rather than at formulating complete theories as to the nature of the processes which may be operating. Such surveys are in fact an essential first step at this stage of our investigations.

One group of tissues that are in some ways particularly suitable for such preliminary investigations are those of the developing Drosophila embryo and pupa. In this animal we find small groups of cells, of a size suitable for electron microscopy, in which contiguous elements may be differentiating in quite different directions, so that we have an opportunity easily to compare their ultramicroscopic morphology. Further, there are in this form a wealth of genetic variants in which we can study the morphological changes produced by mutant genes.

One developing system that has recently been studied in some detail is the compound eye (Waddington and Perry). Within a period of some 72 hours from the beginning of pupation, the essential cellular features of the eye become definitely established. The material was investigated in thin sections after fixing in 1% osmic acid. Adequate fixation depends very largely on the rapidity with which the osmic can be brought in contact with the still living tissues.

The structure of the fully formed eye consists of a closely packed array of hexagonal ommatidia (Fig. 1). Around each six-sided ommatidium there are three minute hairs, each of which is connected to a nerve, the whole hair-nerve complex involving a system of four cells. The ommatidium itself has quite a complex architecture. On the external surface is a transparent cornea. Below this is a cup, formed by primary pigment cells, which has at its base a set of four pseudocone cells. The primary pigment cells and pseudocone cells together are responsible for the secretion not only of the cornea but of a transparent material, which lies immediately below it and fills the cup formed by the pigment cells. Beneath the pseudocone cells are eight retinulae.

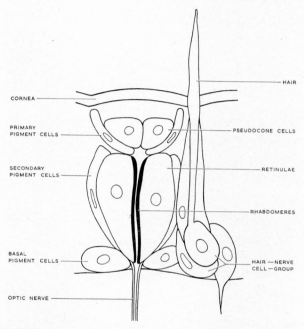

Fig. 1. Diagrammatic radial section of an ommatidium in the mid-pupal stage.

These are large, elongated cells, arranged rather like the segments of an orange. In each group there are six main retinulae all of which have their nucleus at approximately the same level, towards the outer end of the cell. The seventh retinula is also a large cell, with a nucleus which lies at a somewhat deeper level than the other six. The nucleus of the eighth retinula originally lies at a still deeper level, and this cell eventually becomes very small and almost completely disappears. Along the major axis of each of the seven main retinulae there appears a transparent rod, known as the rhabdomere. These rhabdomeres, and the transparent material which appears between them, form as it were, the central strand of pith in the orange. Forming a sheath round the whole ommatidium there are a group of secondary pigment cells, while around the innermost tip of the ommatidium there is a further group of basal pigment cells, between which the optic nerves from the ommatidia extend through to the brain. The hair nerve cell groups originally also lie near the basement membrane but as the ommatidia extend in length, they get left behind so that in the later stages they lie as a relatively high level in relation to the retinulae.

In the late larva, which is just ready to undergo pupation there is

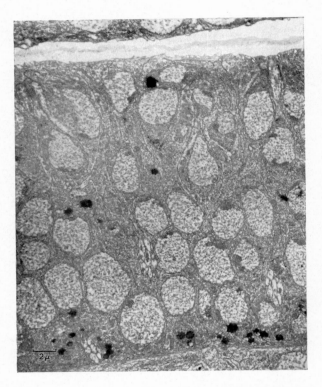

Fig. 2. Radial section, optic disc of late larva. Peripodial membrane above, closely packed presumptive ommatidial cells forming a thick layer below.   (× 6000)

very little sign of this elaborate cellular architecture (Fig. 2). The cells of the optic disc are very closely packed together. In EM sections the most striking feature is the density of the cell membranes. These take a very sinuous and contorted course, indicating that the cells are elaborately interdigitated with one another. The cytoplasm is meagre, and only a few mitochondria and other cytoplasmic structures can be found. The nuclei are large and the nuclear membranes ill-defined. Nucleoli are present but usually rather small.

At first sight there appears to be no orderly arrangement of the cells at this stage. This is, however, misleading. It is known from light microscope studies that near the surface of the disc the cells form small rosettes, and these can also be found in EM sections taken parallel to the surface. It is not easy to determine with either instrument exactly how many cells are involved in these groups. The cells are elongated and spindle shaped, and not all of them have their thickest region in the same transverse plane. Groups of five or six cells making up a rosette in a transverse section therefore probably represent only a fraction of the total group. In some sections one comes across what is clearly a group of the basal tips of these elongated cells. These are commonly in groups of eight, which appear to correspond to the eight retinulae. It is not clear whether the future pseudocone cells are already attached to the distal end of these groups or not. Since these groups of basal tips appear not only towards the basement membrane of the optic disc but also within its thickness, it is clear that the groups of cells must be displaced relative to one another in the proximo-distal direction. Details of the structure of the ommatidia forming layer are therefore rather complex, and in view of the complicated interdigitation of the cells it will be difficult to work them out in detail.

The essential fact from our present point of view is the paucity of cytoplasmic structure and the similarity between all the types of cells in the optic disc. By 24 hours later, and still more by 48, this uniformity of appearance has been replaced by very striking structural differentiation. The full details of this must be described elsewhere but here some of the most salient points may be mentioned to illustrate the phenomenon with which we are faced in the study of differentiation.

Perhaps the most spectacular feature in the submicroscopic morphology of the developed ommatidium is the meshwork forming the rhabdomeres (Figs. 3 and 4). The structure of this organ in the adult insect has been described by Fernandez-Moran (1956) and Danneel and Zeutzschel (1957). There is a closely packed mass of small tubular elements, which have rather definite arrangements in the different retinulae. The retinulae themselves are arranged with remarkable regularity; the seven cells protruding into the central cavity make a very definite pattern which is repeated precisely over large areas. One might have thought that these protrusions were the result of chance and would vary from ommatidium to ommatidium, but that is not the case; the pattern remains exactly the same over a large number of retinulae and may then suddenly change to another pattern, which again is repeated over a considerable area, and within each ommatidium the orientation of tubules within the rhabdomeres is then quite orderly and by no means a matter purely of chance.

Figure 3 shows an early stage in the development of the retinulae.

The vesicles from which the rhabdomeres will be formed are just beginning to accumulate in the region where the retinula cells come in contact with one another. Very similar vesicles can be seen within the cytoplasm of the retinulae, and the appearances suggest that the vesicles are not formed *in situ* in the position of the rhabdomeres but are produced at first in the cytoplasm and then migrate to their peripheral location. In fact, in several sections, groups of vesicles are found near

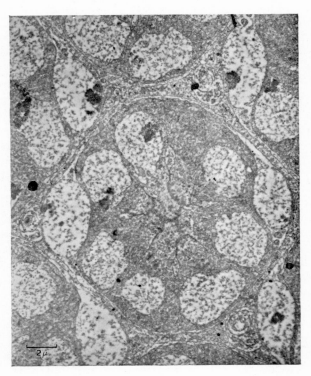

Fig. 3. Transverse section of ommatidium, 48-hour pupa. The nuclei of the seven main retinula cells are clearly seen, forming a group which is cloted by two secondary pigment cells, whose nuclei lie one on each side to right and left. Three "hair-nerve" groups can be seen (one at left, one above right and one lower right). In the centre of the ommatidium vesicles are just beginning to collect at the junction of the seven retinulae: these will later form the rhabdomeres. ($\times$ 7800)

the nuclear membrane, which is often very much folded and contorted in the region of the nucleus which faces towards the rhabdomere edge of the cell. It would not be difficult to imagine that the rhabdomere vesicles are being initiated in the close proximity of the nuclear membrane, but it would of course require something more than purely morphological evidence to demonstrate this conclusively. That the retinulae nuclei are very active would, however, appear to be indicated by the presence of large nucleoli. These always lie against the nuclear membrane, usually towards the periphery of the cell, that is, on the side away from the central rhabdomeres. In the immediate neighbourhood of the nucleoli, the nuclear membrane becomes difficult to distinguish, and one often has the impression that it is actually absent, so that the nucleolus material can escape freely into the cytoplasm.

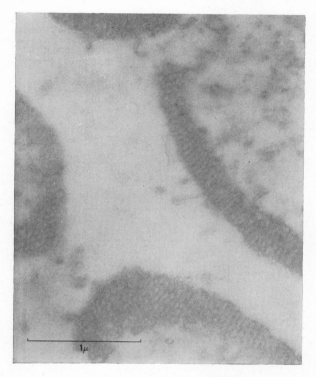

Fig. 4. Rhabdomeres of 4 retinula cells, transverse section, 72-hour pupa. ($\times$ 60,000)

Another striking feature of the differentiation is the formation of the cornea by the so-called primary pigment cells and the pseudocone cells (Fig. 5). These cells possess an elaborate endoplasmic reticulum and are also full of well developed mitochondria. At the distal surface of both the pseudocone cells and the primary pigment cells, the cell forms a series of small peg-like protrusions, and it is, apparently, from these that the secretion of the corneal material takes place. In Fig. 5, which is cut tangentially to the corneal surface, one can see that the cornea has a lamellar structure which, in its lowest layer, breaks down into a series of wisps of newly secreted material. Immediately below, the peg-like protrusions are seen arranged close together over the surface of the cell: in the central (deepest) region of the section these are uniting together into strands which merge, at a still deeper level, with the general cytoplasm of the pseudocone cells.

The hair nerve groups are also very striking features in the architecture of the developing eye disc. These cells, which form a group of the kind common in the insect integument and discussed in its various

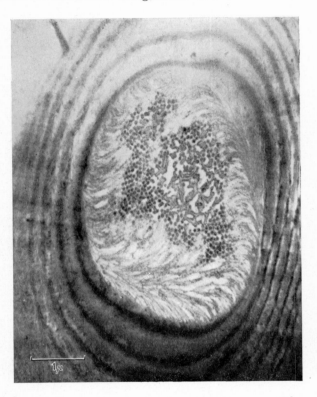

Fig. 5. Transverse section through cornea and underlying pseudocone cells, showing the protrusions from the cell surface by which the cornea is secreted. 72-hour pupa. (× 24,000)

forms by Henke (1953), are closely wrapped round one another so that their detailed arrangement is difficult to disentangle in the short series of serial sections which is all one can hope for with the electron microscope. They are remarkable for the great development of the endoplasmic reticulum in one of them, which is presumably the cell from which the hair itself is secreted (Fig. 6). This cell also contains large mitochondria. In other cells of the group the nucleus has a very characteristic appearance, being surrounded by an extremely dense region which suggests that activity is proceeding around the whole circumference.

These few examples are enough to give some indication of the wealth of submicroscopic structure that is involved in the differentiation of an organ like the insect eye, in which cells of several different types are lying in immediate contact with one another. Somehow or other, these

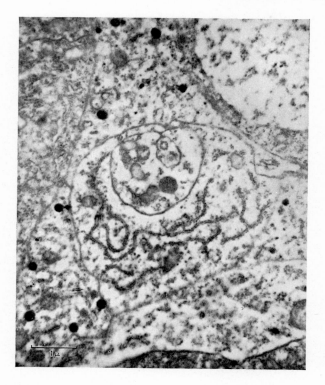

Fig. 6. Transverse section through hair-nerve cell group, distal to the level of the nuclei. The outer cells are seen to be wrapped round the innermost, which is presumably the nerve fibre. Above and to bottom left are secondary pigment cells containing pigment granules. Retinula cells to left and right above. 48-hour pupa.
($\times$ 24,000)

different cytoplasmic patterns must be produced by, and must themselves presumably also influence, the nuclear genes. We should be able to obtain further insight into these relations by the study of mutant strains in which particular genes lead to abnormalities in differentiation. As yet, only a few first steps have been taken in this direction, and until the normal development sequences are more thoroughly worked out, it will be hazardous to try to interpret the appearances that one finds in the mutant individuals. As an example of the type of phenomenon that one may find, Fig. 7 shows a section through an eye of *roughoid-g*. It is clear that some ommatidia contain abnormal numbers of retinulae, and further that an unusually large proportion of the area of the section is occupied by cells belonging to the hair nerve groups. In other sections one can see what appear to be abnormal

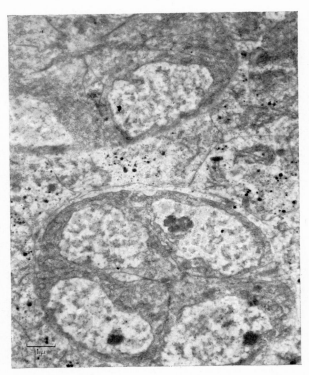

Fig. 7. Transverse section through ommatidia of *roughoid-g*. Note variation in the number of retinula cells and the greatly expanded inter-ommatidial areas. 48-hour pupa. (× 15,000)

cytoplasmic structures in retinular cells, and again reduced numbers of pseudocone cells whose endoplasmic reticulum appears to be abnormal. Another mutant eye we have studied is that of *split*. This is characterized by an oversecretion of the clear material that fills the primary pigment cup under the cornea (Pilkington, 1943). There appears to be an abnormal cytoplasmic structure in the cytoplasm of these cells at the 48-hour stage. Further work will be necessary before one can be certain as to the reality of these abnormalities, and their relation to the more crude morphological changes that have been detected with the light microscope. They are shown here as indications of the *type* of evidence which the electron microscope may make available and which will have to be incorporated in any full understanding of gene action and cellular physiology. They make it clear that there is a new realm of structure open to investigation. It will presumably offer very many problems, but it is perhaps not unduly optimistic to suppose that it may also suggest one or two answers to some of the questions with which the cell physiologist is faced.

References

1. Bellairs, R. *J. Embryol. Exp. Morph.* **6**, (1958).
2. Danneel, R. and Zeutzschel, B. *Z. Naturf.* **12**b, 580 (1957).
3. Fernandez-Moran, H. *Nature, Lond.* **177**, 742 (1956).
4. Henke, K. *J. Embryol. Exp. Morph.* **1**, 217 (1953).
5. Lehmann, F. E. *Naturwissenschaften* **43**, 289 (1956).
6. Pilkington, R. W. *Proc. Zool. Soc. Lond.* A, **111**, 199 (1942).
7. Selman, G. G. and Waddington, C. H. (unpublished) (1958).
8. Waddington, C. H. and Perry, M. M. (unpublished).
9. Weiss, P. *J. Cell. Comp. Physiol.* **49** suppl. 1, 105 (1957).
10. Yamada, T. (unpublished lecture) (1957).

THE ORIGIN OF THE NUCLEUS AFTER MITOTIC CELL DIVISION

J. Boss

Department of Physiology, University of Bristol

At the stage of mitosis known as anaphase there are two groups of daughter chromosomes receding from each other. At the next stage, telophase, there is, instead of each chromosome group, a newly formed nucleus, which will gradually change until it has assumed the appearance characteristic of the interphase between mitotic divisions. In a typical somatic cell of a vertebrate the transition from anaphase to interphase is marked by *three stages* in the transformation of the nucleus.

At first, in each daughter group, the chromosomes come very close together at the end of anaphase and, as they separate again, they are seen to be enclosed by a distinct nucleocytoplasmic boundary, the nuclear membrane. Thus a nucleus is formed, in which there are chromosomes and material in the interstices between them. The chromosomes contain, or consist of, desoxyribonucleo-histone and are Feulgen-positive; they stain deeply with haematoxylin and basic dyes. The interstitial material is, at this stage, free of desoxyribonucleic acid, is therefore Feulgen-negative, and stains weakly with haematoxylin and basic dyes. For convenience, the terms *chromatin* and *chromatinic* will be applied in this chapter to material containing desoxyribonucleic acid (DNA) and staining as the chromosomes do, while *non-chromatinic* will describe material with staining properties indicating the absence of that substance. In the young daughter nuclei, which are the subject of this discussion, chromatin is invariably a component either of the chromosomes or of structures readily seen to have originated from them.

The second stage of nuclear reconstruction is marked by a continuing increase in the separation of the chromosomes and an uncoiling of the threads of which they are composed.

Lastly, the nucleoli become large enough to be easily visible and develop into smoothly rounded bodies, while the chromatin becomes so scattered through the nucleus that only parts of it remain distinguishable as discrete lumps and threads.

This chapter is concerned with the origin of the non-chromatinic material in the first of these three stages. I have already described in detail the first and second stages as they are seen in living and fixed tissue cells of the adult newt *Triturus cristatus* [6], and the discussion presented in the account that follows is based on a comparison of those results with the published findings of other workers.

Types of Chromosomal Vesicle

The term *chromosomal vesicle* has been used to describe structures of a number of kinds, falling into two main groups, chromatinic and non-chromatinic. The former are spaces within the chromatinic material of the chromosomes, while the latter are structures in which chromosomes lie; that is to say, the naming of each type, in this chapter, will depend on the character of the wall of the vesicle.

Chromatinic vesicles are themselves of two kinds. Sharp [49], Kater [25, 26] and Fell and Hughes [16] have described small vacuoles appearing in chromosomes late in mitotic division. Such vacuoles might, as Kaufman [27] has pointed out, be a fixation artifact, a view whith Sharp [50] himself later concurred; and Nebel has suggested how a coiled thread could be converted to a vacuolated column by fixation [39] or appear vacuolated as the result of optical conditions [40]. A second type of chromatinic vesicle is exemplified by those found by Moore [38] in the cleaving blastomeres of hybrid frogs. In these cells the chromosomes in telophase become chromatinic sacs, and seem similar to the hollow chromosomes shown in Kurosumi's electron micrographs of sections of sea-urchin blastomeres [31]. It would be convenient to regard these hollow chromosomes as normal helices in which the central space is distinguishable, but Moore found that the vesicles were fewer in number than the chromosomes from which they were formed, and some examples of the formation of chromatinic vesicles of this second type therefore remain unexplained.

A *non-chromatinic vesicle* is a compartment containing a chromosome or its chromatinic part. It may be noted that chromosomes within non-chromatinic vesicles may themselves contain chromatinic vesicles of the types just described.

Non-Chromatinic Vesicles in Early Telophase

Wilson [57] and Sharp [50] considered that the telophase nucleus might, in some tissues, form by the fusion of chromosomal vesicles, while Hughes [21] suggests that the process is general in mitosis. Now it is quite certain that in cells of many types this is not true if chromosomal vesicles of the chromatinic type are meant. In newts, for example, telophase in somatic cells begins as the separation of temporarily aggregated chromosomes, non-chromatinic material appearing between them and not within them. This is clear from Flemming's account of the tadpole [17], and no vacuoles visible with the light microscope ever appear within the chromosomes in cultures of tissues from the adult newt [5, 6]. On the other hand, if the telophase nucleus is considered as a group of non-chromatinic chromosomal vesicles, a large number of apparently different modes of nuclear reconstruction can be regarded as varieties of a single process; even some instances of the supposed fusion of chromatinic vesicles can be accounted for in this way, as will be shown when reference is made to the work of Richards.

If a chromosome, or a fragment of one, remains separate from the rest of the group it will, in telophase, be found to lie within a non-chromatinic vesicle which, with its contained chromatin, forms a micronucleus. Thus micronuclei form as the result of agents, such as mustard gas [e.g. 22] or ionizing radiations, which disrupt the mitotic figure.

In the cleavage divisions of *Cryptobranchus* [51] and *Cyclops* [20] and in spermatogonia of the grasshopper *Aularches* [56], chromosomal vesicles are apposed to form a single daughter nucleus, but they remain throughout interphase in *Cyclops* and *Aularches* as nuclear compartments, each containing a chromosome, and probably do so in *Cryptobranchus* also. The spermatogonia of some orthopteran insects show a further degree of fusion of the vesicles. In *Phrynotettix magnus* [44, 55] the walls of the nuclear compartments break down in interphase but reappear in prophase, while in *Stethophyma grossum* [24] the compartments, once fused after telophase, do not again become distinct before the next mitosis. In spermatogonia of the bush-cricket *Pholidoptera griseoaptera* [56], all the vesicles fuse except that of the X-chromosome, which remains in a distinct compartment throughout interphase.

When chromosomes are close together within their daughter group at the end of anaphase, discrete vesicles may not form around the chromosomes, as the non-chromatinic material forms a continuous mass from the first. Thus, for example, the chromosomes of somatic cells of the newt become enclosed within a single Feulgen-negative mass at the beginning of telophase, this mass, with the enclosed chromosomes, being the newly formed daughter nucleus [5, 6]. Although, in such nuclei, discrete vesicles are not formed, the relatedness of the non-chromatinic material of the nucleus to the chromosomes is shown by the shape of the nucleus, which roughly corresponds to the outline of the daughter chromosome group at the end of anaphase. The shape of the anaphase group may be preserved throughout interphase, as for example, in some blood cells of the rat [41].

Just as the non-chromatinic vesicles in telophase exhibit behaviour which can be arranged in a graded scale, ranging from persistent separation to initial union, so their chromatinic contents may also vary in the changes that they undergo. At one extreme are chromosomes that remain compact and distinct throughout interphase. These are found, for example, in flagellate protozoa [13, 14, 19] and orthopteran spermatogonia [56]. At the other extreme are those cells, common in the tissues of vertebrates, in which the chromosomes become more and more diffuse, presumably by their fraying into submicroscopic threads [cf. 6], until only a small amount of discrete chromatin is resolvable by the light microscope and the nucleus is elsewhere diffusely Feulgen-positive throughout its extent (except in the nucleoli); the diffuse background to the discrete chromatin may indeed be seen as very fine threads but these are not visible in life, notwithstanding the high resolution obtainable by phasecontrast microscopy [42, 53, 6]. Intermediate between these extremes are nuclei in which the chromatin is broken up into discrete threads and grains (chromonemata and chromocentres) and in which these easily resolvable structures account for the total darkly staining material, the spaces between them being free of chromatin. This is the degree of dispersion reached in interphase in blastomeres of the minnow *Fundulus heteroclitus*, as described by Richards [46]. Richards' account of cleavage in the minnow has become the *locus classicus* for chromosomal vesicles. In the blastomeres of the minnow each chromosome becomes surrounded by a non-chromatinic vesicle, and the vesicles aggregate to form a compartmented daughter nucleus. Some of the dispersing chromatin in each vesicle takes up a peripheral position so that each compartment comes to be discontinuously lined by chromatinic threads and grains. "When the nucleus begins the next division the new chromosomes are produced endogenously, each within the substance of one of the old vesicles." That the vesicles are non-chromatinic is clear, but the point is worth making, as later writers have sometimes referred to this work as though the chromosomal vesicles described by Richards were simply expansions of darkly staining chromosomal material, instead of lightly staining bodies within which the material of the chromosomes underwent structural change.

The discontinuous chromatinic lining of the vesicles in the blastomeres of the minnow suggests that chromatin tends to take up a peripheral position in a nucleus or nuclear compartment. In which species and tissues this tendency is observable is a matter on which I have not the knowledge to generalize, but studies on cultures of tissues of the newt *Triturus cristatus* lead to the following conclusions. First, there may be no tendency for chromatin to take up a peripheral position during nuclear reconstruction or interphase. Secondly, certain types of fixation can move chromatin so that more comes to be precipitated against the nuclear membrane. Thirdly, chromatin often comes into a peripheral position in prophase, at the end of which the chromosomes can form a hollow basket, as described and depicted by Fleming [17].

NUCLEOPLASM

For the purpose of the present discussion, the word *nucleoplasm* will be used for the non-chromatinic material of the nucleus, excluding the nucleoli and all particulate matter visible by the light microscope, such as granules containing ribonucleoprotein.

From what has already been said, it can be inferred that the nucleoplasm in telophase is composed of the non-chromatinic vesicles of the daughter group of chromosomes. The vesicles themselves may be evident before telophase, as in Bragg's study of mitosis in rapidly developing early embryos of the toad *Bufo cognatus* [7]. In these embryos each anaphase chromosome appears to consist of a chromatinic core and a non-chromatinic sheath. The sheaths swell to form vesicles which fuse, while the cores give rise to the chromatin of telophase and interphase. Such a sheath is in fact what some writers have called the matrix of the chromosome. Cleveland [14] has described how in *Holomastigotoides tusitala*, a flagellate intestinal symbiont of termites, the nucleoplasm provides the material for the chromosomal matrices of metaphase and anaphase.

Thus it becomes easy to conceive a cycle in which nucleoplasm is alternately matricial and nuclear in form, but it is not to be thought that the whole nucleoplasm corresponds in all cells to the matrix of the mitotic chromosomes. In the flagellate already mentioned Cleveland [13, 14] observed that some of the nucleoplasm does not form chromosomal matrices, but is shed into the cytoplasm in prophase. Furthermore, Reuter [45, pp. 29-30 and Fig. 29] described a presynaptic stage of spermatogenesis in the hemipteran insect *Alydus calcaratus* in which each chromosome is enclosed within a vesicle, but the vesicles are so contained within the nucleus that there is an extravesicular region of the nucleoplasm in which they lie.

THE CHROMOSOMAL MATRIX

The term *chromosomal matrix* has been used in at least three senses. The most intimate type of matrix would consist of desoxyribonucleic acid or its histone salt, and a sheath of such material seems to be postulated by Serra [47, 48]. The variable "nucleic acid charge" reported by workers such as Koller [30], and the "free" desoxyribonucleic acid which Darlington [15] considers necessary for coiling, can also be considered as forming a matrix of this kind. For the present purpose no more need be said about it.

A second type of matrix is that which does not contain desoxyribonucleic acid, but which is so intimately incorporated into the chromosome or so thinly applied to it that there is no visible sheath such as that discussed in the preceding section. In this category is included the charge of ribonucleoprotein [18, 28, 23] and lipid [35] on mitotic chromosomes. (It is probable that Jacobson and Webb [23], in localizing ribonucleoprotein, and La Cour, Chayen and Gahan [36], in localizing lipid, were in fact observing different aspects of the behaviour and composition of similar substances; ribonucleic acid can be associated with a lipoprotein.)

Thirdly, there may be an obvious sheath, readily visible with the light microscope, around a chromosome. The behaviour of this type of matrix has already been referred to. A matrix of this third kind does not seem to be merely an enlarged form of one of the second type, since Jacobson and Webb found that all or most of the charge of ribonucleoprotein was lost from the chromosomes in anaphase, and La Cour and Chayen noted a corresponding behaviour of chromosomal lipid. Nevertheless, in fibroblasts of the newt, ribonucleoprotein continues to be shed from the chromosomes in early telophase and appears in the nucleoplasm [4]. It is possible that some part of the ribonucleoprotein, not shed in anaphase, is a constituent of the non-chromatinic vesicle or nucleoplasm. As far as I know, the distribution of ribonucleoprotein in cells with a thick, readily visible chromosomal matrix has not been examined.

The Nuclear Membrane

The walls of chromatinic chromosomal vesicles differ in structure from nuclear membranes, as Kurosumi's electron micrographs show [31]. On the other hand, if non-chromatinic vesicles are equivalent both to chromosomal matrices of the thick, readily visible type and to the telophase nucleoplasm, the boundaries of these vesicles and of such chromosomal matrices should resemble the telophase nuclear membrane in structure. Consequently, one of the most useful tests of the general view expressed in this chapter would be an electron micrographic examination of anaphase and telophase in cells, such as *Holomastigotoides*, in which each mitotic chromosome has an obvious matrix forming a broad halo round it. Here again, I am not aware that such an investigation has been published.

The nuclear membrane is a submicroscopically oriented structure, as shown by Baud [3] and Callan [8] and discussed by Soudek and Beneš [52], and after fixation has a characteristic structure demonstrable by the electron microscope: see, for example, the publications of Callan [9], Callan and Tomlin [10], Kautz and de Marsh [29],

Watson [54] and Pappas [43]. On the other hand, the structure appears to be mechanically labile; Chambers [11] found that the nuclei of echinoderm eggs could be pinched into droplets which would re-unite to permit subsequently normal embryonic development, while Battaglia [1] has recorded spontaneous fusion of nuclei in *Sambucus ebulus*, an elder, and Lewis [37] has noted the spontaneous fragmentation of nuclei in animal cells in tissue culture.

There is, in fact, no contradiction between the two views of the nuclear membrane: it is a complex structure as well as a labile boundary. At the interface between two liquids the conditions differ from those obtaining within either phase and, if both contain electrolytes some of which carry more than one charge per molecule, the interface will bear charges on each aspect. These in turn may cause the further aggregation of solutes in a regular arrangement, so that there arises between the two solutions a polarized boundary membrane of complex structure, the molecules of which are perhaps not stationary in unfixed material but are in equilibrium with those in the solutions on either side.

Three consequences follow from this view of the nuclear membrane. First, it should be disorganized by surface-active ions, but, since the ions must take up positions at a charged surface, it is to be expected that anions and cations would be unequally effective. In fact, as Baud [2] has shown, anionic detergents applied externally are far more effective than cationic in disorganizing the nuclear membrane; this suggests that the membrane is positively charged on its cytoplasmic side. Secondly, at the time when the nuclear membrane forms (that is, if there has been no persistent vesicle or equivalent matrix), the cytoplasm and nucleoplasm must be immiscible. Thirdly, if they later become miscible, rupture of the membrane must lead to irreversible disorganization of the cell; Chambers and Fell [12] found that cells in interphase could be killed by puncturing the nuclear membrane. The change to miscibility may be due to the hydration of the nucleus which, Kuwada and Nakamura [32, 33, 34] have proposed, takes place during telophase.

REFERENCES

1. BATTAGLIA, E. *Nuovo G. Bot. Ital.* n.s. **54**, 724 (1947).
2. BAUD, CH. -A. *C. R. Soc. Biol. Paris* **142**, 181 (1948).
3. BAUD, CH. -A. *Exp. Cell Res.* Suppl. **1**, 47 (1949).
4. Boss, J. *Exp. Cell Res.* **8**, 181 (1955).
5. Boss, J. *Proc. XV Int. Congr. Zool. London* § 9, paper 9 (1958).
6. Boss, J. *Exp. Cell Res.* **18**, 197 (1959).
7. BRAGG, A. N. *Trans. Amer. Micr. Soc.* **58**, 357 (1939).
8. CALLAN, H. G. *Exp. Cell Res.* Suppl I, **48**, (1949).
9. CALLAN, H. G. *Symp. Soc. Exp. Biol.* **6**, 243 (1951).

10. CALLAN, H. G. and TOMLIN, S. G. *Proc. Roy. Soc.* B **137**, 367 (1950).
11. CHAMBERS, R. *Biol. Bull.* **41**, 318 (1921).
12. CHAMBERS, R. and FELL, H. B. *Proc. Roy. Soc.* B **109**, 380 (1931).
13. CLEVELAND, L. R. *Trans. Amer. Phil. Soc.* **39**, 1 (1949).
14. CLEVELAND, L. R. *Trans. Amer. Phil. Soc.* **43**, 809 (1953).
15. DARLINGTON, C. D. *Nature, Lond.* **176**, 1139 (1955).
16. FELL, H. B. and HUGHES, A. F. W. *Quart. J. Micr. Sci.* **90**, 355 (1949).
17. FLEMMING, W. *Arch. Mikr. Anat.* **16**, 302 (1879).
18. FROLOVA, S. L. *J. Hered.* **35**, 235 (1944).
19. GRASSÉ, P. -P. *Proc. XV Int. Congr. Zool. London*, § 6, paper 52 (1958).
20. HEBERER, G. *Z. Mikr. -Anat. Forsch.* **10**, 169 (1927).
21. HUGHES, A. (F. W.) "The Mitotic Cycle." p. 32; London (1952).
22. HUGHES, A. F. W. and FELL, H. B. *Quart. J. Micr. Sci.* **90**, 37 (1949).
23. JACOBSON, W. and WEBB, M. *Exp. Cell. Res.* **3**, 163 (1952).
24. JANSSENS, F. A. *Cellule* **34**, 135 (1924).
25. KATER, J. McA. *Z. Zellforsch.* **5**, 263 (1927).
26. KATER, J. McA. *Z. Zellforsch.* **6**, 587 (1927).
27. KAUFMANN, B. P. *Amer. J. Bot.* **13**, 59 (1926).
28. KAUFMANN, B. P., McDONALD, M. and GAY, H. *Nature, Lond.* **162**, 814 (1948).
29. KAUTZ, J. and DE MARSH, Q. B. *Exp. Cell Res.* **8**, 394 (1945).
30. KOLLER, P. C. *Proc. Roy. Soc.* **133**, 313 (1946).
31. KUROSUMI, K. *Protoplasma*, **49**, 116 (1958).
32. KUWADA, Y. and NAKAMURA, T. *Cytologia, Tokio* **10**, 492 (1940).
33. KUWADA, Y. and NAKAMURA, T. *Cytologia, Tokio*, **12**, 14 (1941).
34. KUWADA, Y. and NAKAMURA, T. *Cytologia, Tokio*, **12**, 21 (1941).
35. LA COUR, L. F. and CHAYEN, J. *Exp. Cell Res.* **14**, 462 (1958).
36. LA COUR, L. F., CHAYEN, J. and GAHAN, P. S. *Exp. Cell Res.* **14**, 469 (1958).
37. LEWIS, W. H. *Anat. Rec.* **97**, 433 (1947).
38. MOORE, B. C. *J. Morphol.* **101**, 209 (1957).
39. NEBEL, B. R. *Bot. Rev.* **5**, 563 (1939).
40. NEBEL, B. R. *Symp. Quant. Biol.* **9**, 7 (1941).
41. OHNO, S., KINOSITA, R. and WARD, J. P. *Naturwissenschaften* **41**, 288 (1954).
42. OSTERBERG, H. *J. Opt. Soc. Amer.* **40**, 295 (1950).
43. PAPPAS, G. D. *J. Biophys. Biochem. Cytol.* **2**, (Suppl.), 431 (1956).
44. PINNEY, E. *Kans. Univ. Sci. Bull.* **4**, 309 (1908).
45. REUTER, E. *Acta Zool. Fenn.* **9**, 1 (1930).
46. RICHARDS, A. *Biol. Bull.* **32**, 249 (1917).
47. SERRA, J. A. *Symp. Quant. Biol.* **12**, 192 (1947).
48. SERRA, J. A. *Exp. Cell Res.* Suppl. I, 111 (1949).
49. SHARP, L. W. *Cellule* **29**, 297 (1913).
50. SHARP, L. W. "Introduction to Cytology." (3rd edn.) pp. 140-1; New York, (1934).
51. SMITH, B. G. *J. Morphol.* **47**, 89 (1929).
52. SOUDEK, D. and BENEŠ, L. *Českoslov. Biol.* **4**, 416 (1955).
53. VAN DUYN, C. JR. *Microscope* **11**, 196 (1956).
54. WATSON, M. L., *J. Biophys. Biochem. Cytol.* **1**, 257 (1955).
55. WENRICH, D. H. *Bull. Mus. Comp. Zool. Harv.* **60**, 55 (1916).
56. WHITE, M. J. D. "The Chromosomes." Fig. 1., p. 3; London, (1937).
57. WILSON, E. B. "The Cell in Development and Heredity." (3rd edn.); Macmillan, New York (1928).

LABELLED ANTIBODIES IN THE STUDY OF DIFFERENTIATION

R. M. CLAYTON

Institute of Animal Genetics, Edinburgh

INTRODUCTION

THE precise localization of specific macromolecular substances can give valuable information on their synthesis, distribution and behaviour, and is important in studies of cell physiology and differentiation. There are a number of methods that can show such localization with an accuracy subject both to the precision of the methods of recognition and to the errors involved in the methods of preparation: chemical or immunological tests of cellular fractions obtained by differential centrifugation; autoradiography; microspectrophotometry; and histochemical methods with reagents specific for a particular chemical grouping or dependent upon a particular enzyme action. A fifth method, the use of labelled antibodies for histochemical localization, has special advantages. Antibodies are exceedingly precise reagents and, by using fluorescein isocyanate or some other fluorochrome as label, it is possible to determine the site of the antigen in the cell, while with a radioactive label semi-quantitative estimates are possible. With layer techniques, minute traces of antigen may be detected. With two fluorochromes the relative positions and simultaneous changes in two antigens may be determined. Finally, internal evidence suggests that there is little or no diffusion of antigens in the methods employed.

The main technical difficulties result from non-specific deposition of labelled antibody, which may be eliminated or checked by controls; from cross-reactions of the antisera, which are not always easy to eliminate in the case of antibodies to normal tissue constituents, but present less trouble when localizing pathogens in a host; and from the choice of methods of fixation. The subject is reviewed by Coons [22].

Coons et al. [18, 19], having shown that fluorescein isocyanate is a suitable label by virtue of its brightness, colour and firm conjugation to protein, were the first to use labelled antisera as histochemical reagents, and Marshall [56] first applied Coon's method to normal tissue constituents.

The main subsequent additions to the basic technique are: the use of sandwich techniques to localize antibody; layer techniques for enhancement [see 22 for references]; a method of studying tissues *en bloc* suitable for insoluble antigens [31], and the use of radio-iodinated antisera or several fluorochromes simultaneously in order to meet the requirements of certain investigations [4, 4a, 11, 12, 13, 80]. Fluorescent antisera can give precise intracellular localizations; radioactive antisera (generally I^{131}) are relatively poor in this respect, but are probably more sensitive.

Coons' basic technique is in general, however, followed closely although some workers have introduced modifications or raised problems from time to time. The technique, described in detail in the earlier papers, together with the nature of the essential controls is summarized by Coons in his 1956 review, as are some of the variations in the procedure for producing or fixing the sections [see also 23b and 82]. The bulk of the localizations made by the immunohistological method are of pathogens or of injected foreign proteins rather than of antigens native to the tissue. However, a number of these have been examined, most of them since 1956, and, since some special problems are raised, it is perhaps worth summarizing the main technical points. Many of these have already been discussed by Coons and have been merely listed below. It is not the purpose of this review to refer to all the recent work using labelled antibodies. All work on pathogens (with the exception of viruses, which are a potential source of information on protein synthesis) and all clinical studies that are not relevant to tissue differentiation are excluded from this review. Cruickshank lists some recent references [26].

TECHNIQUE

1. *The antiserum*

Many workers have pointed out that it is an advantage to employ an antiserum as specific as possible to eliminate cross-reactions which may be difficult to interpret. Absorption of an antiserum will frequently eliminate many of these, but some difficulties may still be encountered. The literature on immunological analysis of tissue antigens (*in vitro*) indicates that in the process of eliminating cross-reactions, the titre of an antiserum may drop to a uselessly low value and that the relative purification of antigens may not always eliminate cross-reactions.

2. *Preparation of material*

Freezing and Cutting. One new point only need perhaps be made and that is on the freezing of delicate embryonic structures. Early embryos are extremely fragile and consist largely of water; their cells may disrupt on exposure to air, while an embryo plunged into a freezing mixture in a drop of saline is crushed beyond recognition. They can be preserved, however, in the following manner:

Mouse embryos up to ten days' gestation may be frozen in the embryonic membranes or pieces of uterus. Chick and amphibian embryos may be transferred into

2% agar or 10% gelatin which is lukewarm and about to set. The embryo and surrounding gel may be cut out taking care not to compress the gel towards the embryo, which causes disruption of the cells, and the whole dropped into the freezing liquid (generally isopentane or arcton-6 (ICI) surrounded by a jacket of liquid nitrogen). (Various solvents containing crushed dry ice are often used to freeze tougher tissues, which may be dropped into test tubes.) Thick sections may be cut in a cryostat but it is far easier to freeze-dry the specimen on top of a bed of previously de-gassed low melting point wax *without removal* of the material from the apparatus. A dusting of carbon black on the gel before freeze drying will serve to locate it in the paraffin if necessary.

The principal methods of preparation of sections are summarized by Coons; a few additional points may be made. Wax sections prepared by the Altman-Gersh method should preferably not be floated on water in view of the finding [69] that proteins and carbohydrates may be lost from an unfixed paraffin section. The material may be fresh-frozen and cut in a cryostat or with a chilled knife. The usual method of attachment of the sections by brief thawing through the underside of the slide is suitable for tissues that are structurally tough and compact, and where the antigens under study do not smear themselves over the section more rapidly than re-freezing and/or drying can take place. With all embryonic and some other material this method is not suitable, as a structureless smear forms instantly on raising the temperature to 0°. Instead sections may be attached to gelatin films or gelatinized slides [10], and the gelatin may be briefly thawed through the slide, but the section should not itself thaw. Fresh-frozen sections have been dried with a fan at room temperature, at low temperatures, *in vacuo* or by substitution with alcohol at sub-zero temperatures. The first method is obviously quite unsuitable for embryonic material.

Glycerol treatment (rendering cells permeable to antibody) followed by antibody, formalin fixation, and ordinary embedding and cutting [31] is suitable for antigens insoluble in glycerol-saline.

Fixatives. Several fixatives have been used:—95% ethanol at 37° or at 0°, absolute methanol at 0°, acetone, formalin-dioxane (see Coons, 1956); also osmic acid vapour [3] and some material has been left unfixed [38, 42]. Very little has been done in the way of checking the effect of fixatives on test-tube immunological reactions. Van Doorenmaalen [27] finds that his lens antisera react with alcohol or acetone-treated antigen. However, the antisera, which reacted with iris extracts *in vitro*, localized only in the lens and not in acetone-fixed iris in sections, while retinal localizations were obtained after formalin dioxane, or cold methanol fixation [12, 13]. Viruses can be localized in unfixed material or after acetone fixation, while serum proteins are normally fixed in ethanol. Kaplan finds that one of the two sarcoplasmic antigens he studied fixes well in acetone and the other does not [48]. Finally, the material may in some cases be fixed before cutting and subsequent treatment with anti-sera. This may be done with some polysaccharide antigens (references in Coons [22]) and has been done for chick embryos by Van Doorenmaalen who dehydrated and fixed in cold acetone, and for the amphibian embryo after fixing with alcohol in the cold [14]. Fixation in formol acetic followed by conventional section cutting has also been used [64]. Clearly, every antigen is a special case and a fixative must be found that causes the minimum denaturation or other change of the immunologically reactive grouping, while preventing loss of antigen.

3. *Immunohistological procedures*

1. Direct method with labelled antisera.
2. Sandwich method using an intermediate layer of antigens, followed by labelled antisera and used to detect antibody globulin.

3. Indirect method using an unlabelled antiserum followed by a labelled antiglobulin antiserum. This is formally similar to a Coombs test. The intensity of reaction is enhanced.

These three methods and their proper controls are reviewed and summarized by Coons.

4. Complement has been used as an intermediate layer by Goldwasser and Shephard [39].

5. Double labelling method. Two antisera of different specificities may be coupled each to a different and contrasting fluorochrome which is mixed and applied to the sections. The specificity of each may be tested individually by inhibition tests or by other controls. Clayton [12] used 1-dimethylaminonaphthalene-5-sulphonyl chloride, a fluorochrome developed by Weber [86] which has a firm attachment to protein and fluoresces bright lemon yellow. This fluorochrome has been re-investigated by Redetzki [73] who finds that the antigen antibody reaction *in vitro* is not affected by the coupling of the dye to the antibody; and by Mayersbach [57c]. Nuclear fast red which fluoresces crimson was also found to combine firmly with protein. Some initial confusion was caused by the fact that there are three dyes of this name but entirely different constitutions. The formula quoted from Conn [17] is inappropriate, that of the nuclear fast red in question is a manufacturer's secret; it is not a rhodamine, unlike the third nuclear fast red available in Britain. Uncombined dye may be removed by reprecipitating the protein with ammonium sulphate or electrophoretic dialysis. More recently, rhodamines have been suggested as alternative fluorochromes, one by Silverstein [80] and one, RB200, by Chadwick, McEntegart and Nairn [11]; this is now used by a number of workers and produces an orange-red fluorescence. Hiramoto *et al.* have used another rhodamine with a similar colour [43].

6. Radioactive antisera. These have been used by Pressman and coworkers to localize injected organ antisera to lung and kidney *in vivo* [70, 71] and in estimations of erythrocyte antigens by Boursnell *et al.* [6]. They may also be used [4, 13] as histochemical reagents followed by autoradiography where quantitative estimations may be of interest.

7. Singer [80a] has applied sera labelled with ferritin to fixed sections, afterwards examined in the electron microscope. The persistence of specificity after appropriate fixation is a pre-requisite.

8. Non-specific staining. The elimination or control of non-specific staining is a very important factor and is fully dealt with by Coons [22]. See also [26a, b, 57d].

DIFFERENTIATION

One of the central problems of embryology is the process of differentiation and the nature of the chemical processes underlying it. The changing relations to one another of differentiating cells and the pattern of synthetic changes occurring in cells during ontogeny have been examined by various techniques and in recent years these problems have been approached immunologically, [Reviews 30, 83, 95], both by *in vitro* and *in vivo* methods. Immunohistological localization has so far been applied to embryological problems on a very small scale only. There is, however, a growing body of information on the localization of antigens native to the tissue which is of great interest to embryologist, cytologist and physiologist alike. The antigenic structure of adult tissues and cells defines the end-point to which the embryonic processes of differentiation must be directed. Since some of these adult antigens are sufficiently well defined and localized immunologically, they would be excellent material for the regressive analysis of the embryologist. Of the investigations described below, the studies of the antigenic constitution of the kidney may be regarded as defining such a complex end-point, while the distribution of blood group substances in gut mucosa suggest strongly that an end-point is achieved partly by a divergence during development of related synthetic abilities of similar cells. (This type of event has been indicated by several embryological studies.) Another factor of importance in the maturation of the fully differentiated tissue is the acquisition of new synthetic abilities, and here, data on the localization of antigens in actively secreting or synthesizing cells may provide information on the sites of synthesis within the cell. The data may be compared with that obtained from centrifuged cell fractions and radioactive tracer procedures. The relevant experiments described below are those giving some information in the intracellular localization and the synthesis of protein hormones, serum proteins and virus proteins. Some data on serum proteins in disease have been included here. The investigations on muscle and lens are next described; while these are characteristic structures in the mature animal, the antigens have been localized intracellularly and they have been followed in the developing embryo, so that some part of the history of their differentiation from the earliest stages of their detection to the final structure is available. Finally, some problems which are peculiarly those of embryology and cell biology will be described, namely induction and normal cell permeability (there are data on abnormal tissue permeability in certain clinical conditions; these have not been reviewed here).

The kidney

Pressman and co-workers have shown that a radio-iodinated anti-kidney serum localizes in the glomerulus when injected *in vivo* [70, 71]; Hill and Cruickshank, and Cruickshank and Hill labelled antisera against lung, whole kidney, and kidney glomeruli with fluorescein and localized the antigens reacting with these antisera in fresh-frozen preparations [24, 42]. The glomerular, and to a lesser extent, the lung antisera stained the basement membrane of the glomerulus and tubule and both reticulum and basement membrane in other tissues studied. Anti-kidney on the other hand localized in the basement membrane of the glomerulus and the tubule, and in tubule cytoplasm. The staining of the basement membrane could be removed by absorption of the antisera with lung or some other source of basement membrane and an even staining of the tubular cytoplasm remained. Although the nephrotoxic antisera to basement membrane obtained by these workers is able to detect all basement membrane antigens in sectioned tissues, the nephrotoxic antisera of Pressman and co-workers localized only in the glomerulus. Mellors *et al.* [59], and Ortega and Mellors [68] have shown, by using a fluorescent antiglobulin to detect the site of action of injected nephrotoxic antisera, that the *in vivo* localization was exclusively in the basement membrane of the glomerulus (see also [43a]). Localization of basement membrane antigens has thus determined the total distribution in the body of the antigens which for physiological

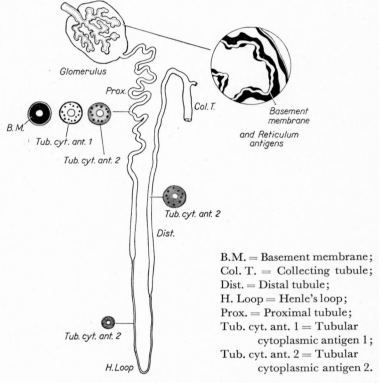

Fig. 1. Distribution of kidney antigens.

reasons are exposed to the action of antisera in a very limited region of the body. Hill and Cruickshank reported that the antiserum which detected basement membrane in reticulum did not stain collagen or cartilage. Goodman, Greenspon and Krakower using *in vitro* methods have come to the conclusion that the basement membrane of the glomerulus contains a high proportion of collagen [40]. Nevertheless, the negative reaction of Cruickshank and Hill is to be expected, since it is exceedingly difficult to produce an antiserum to collagen [85].

Goodman *et al.* also consider that the basement membranes are antigenically complex and in particular that of the glomerular basement membrane contains both tubule and glomerulus specific components, in addition to a collagenous antigen. (See also [50a].) Scott has been able to distinguish between basement membrane and reticulum [78] by the use of fluorescent antisera to the glomerular basement membrane which stained a number of structures which the antisera to reticulum did not. This evidence, supported by mutual blocking experiments suggests that the kidney glomerulus basement membrane contains two related antigens. Scott's sera, like those of Hill and Cruickshank, were negative for collagen. Yagi and Pressman also find several glomerular antigens [97].

Leaving aside the question of the complexity of the basement membrane we may consider the evidence on the cytoplasmic antigens of the kidney. Hill and Cruickshank localized a tubular cytoplasmic antigen with their serum; its distribution in the cell is fairly uniform. Weiler who made fluorescent antiserum to kidney cytoplasmic particles localized cytoplasmic tubular antigens which have an uneven distribution in the proximal convoluted tubule, being mainly situated at the brush border [89]. The same serum stains distal convoluted tubule more evenly and more faintly and Henle's loop evenly and more faintly still. Weiler's antigen and Hill and Cruickshank's antigen are therefore not identical but Hill and Cruickshank's tubule cytoplasm antigen probably includes that of Weiler's. Weiler was able to show by *in vitro* adsorptions with the kidney and tumour antisera followed by histological localizations that there are at least two cytoplasmic antigens, one of which is lost in transforming to a kidney carcinoma. Weiler has since shown that a tubule cytoplasmic particle antigen is lost on tissue culture, while a lung-kidney one persists (89a). Goodman, Greenspon and Krakower report on further cytoplasmic antigens in the glomerulus [40]. Antisera to renin were found by Nairn *et al.* [64a] to localize in the capsule and glomerular tuft. Other data [4b] are not at present in full agreement.

Gut

The distribution of blood group antigens in the gastric and duodenal mucosa has been studied by Glynn, Holborow and Johnson [38]. Their data have been summarized in Fig. 2 and the following points of interest may be noted. The ability of the cells to synthesize the mucopolysaccharide antigen follows a definite and regular pattern, which depends on the region of the gut observed. While the same cells may synthesize A or B antigens and H substance a double labelling experiment [45] shows that Le^a antigens and AB antigens are mutually exclusive in cells. One cell may secrete A, its neighbour Le^a. This competition between one pair of similar substances and lack of it between another pair presents a problem of great interest in the kinetics of cell differentiation. There are equally sharp distinctions in the cells in the glands of the stomach between cells which secrete water soluble blood group substance and those secreting it in an alcohol soluble form.

The presence of mucopolysaccharide in parietal cells is confirmed by Birnbaum and Wolman [5]. Antigenic changes from AB to A or B in some of these cells, probably due to genetic changes in the cells, have been observed [16a].

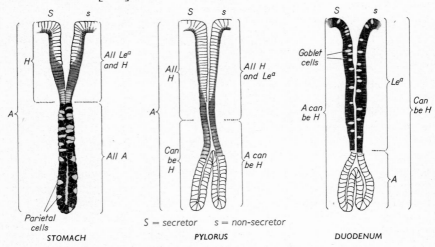

Fig. 2. Distribution of blood group substances in gastric nucosa.

Glynn and Holborow [38a] have since found blood group antigens in several other tissues.

Erythrocytes and leucocytes

Alexander [2] has shown that red cells stain with antibody to species-specific stromal antigens but not with antibody to group A antigen; this difference being almost certainly due to the relative amounts of these antigens present in the red cell envelope. Jankovic and Lincoln [47b] have demonstrated D antigens on leucocytes.

Liver

Weiler [87] has shown that an antiserum to liver cytoplasmic particles will localize in normal liver but not in a hepatoma. This was confirmed by Louis [55], but later Hughes *et al.*, confirmed the localizations but presented evidence that they are non-immunological in character [47].

These workers (47a, 48a, 55a,b) have compared some normal tissues, which bound any labelled sera non-specifically, with the corresponding neoplasms, which do not; they consider that the difference is due to a loss of basic proteins in the neoplasms. However, not all normal tissues do this (see [48a] and particularly [57d]) and neoplasms have been successfully labelled by specific antisera both by Weiler and by Hiramoto et al. (43b, c). Two phenomena are therefore probably involved in some of these studies: changes in specificity and in basicity.

Brain

The serum of rabbits autoimmunized against brain localizes in the myelin sheath of nerves (Buetner et al. [9]).

Paramecium

The genetically controlled antigen responsible for immunologically specific immobilization has been localized by Beale and Kacser [3].

Hormones

Protein hormones, if antigenically pure and endocrinologically active preparations can be made, may be suitable material for immunohistological localization. Marshall [56] has reported on the localization of hog adrenocorticotrophic hormone (ACTH) which he found in the cytoplasm of the basophil cells of the hog pituitary. The nuclei were negative and the cytoplasm appeared to be full of fine densely packed granules. Cruickshank and Curry [25] have examined several protein hormones of human pituitary. They obtained localization of ACTH, gonadotrophin, and thyroid-stimulating hormone (TSH) in cytoplasmic granules of the basophils, and growth hormone in the nuclear membrane and cytoplasm of cells which were unidentified, but not basophils. They were unable, however, to eliminate staining by blocking or adsorption except in the case of growth hormone staining for which could be eliminated by adsorption of the antiserum. They discussed various reasons for the results: the major problem is certainly the difficulty or impossibility of obtaining antisera that are specific for one pituitary hormone at a time. According to the *in vitro* test most absorptions of any of the antisera removed almost all the antibody, but the failure of blocking histologically suggests a tremendous overlap rather than identity of antigens. Marshall did not try any blocking experiments nor did he test the pituitary cross reactions of his antigens but he tested his preparations for hormone activity and showed that normal rabbit serum did not localize in a comparable way. It seems then that the localizations both of Marshall and of Cruickshank and Curry are biological facts, although the precise nature of the antigens being localized is doubtful in every sense except the pharmacological one.

Procarboxypeptidease and chymotrypsin were found by Marshall [57] in the zymogen granules and adjacent cytoplasm of the cells of pancreatic acini. The nucleus and the rest of the cell were negative and the hormone was found in the side of the cell directed towards the lumen. White [92] finds that thyroglobulin is similarly located in the apex of acinar epithelial cells of the thyroid. It was noted that where an acinus was full of colloid the epithelial cells were stained more faintly than where it contained a small quantity [15]. These two findings confirm those made by autoradiography of animals which have been fed I^{131} [28, 53]. White has since extended his observations and found thyroglobulin in the nucleus and cytoplasm of immature cells, and in the cytoplasm of mature cells, especially towards the apex. [94a].

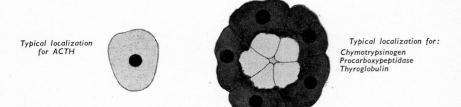

Fig. 3. Synthesis of hormones.

White [92], Hiramoto et al. [43] and Buetner et al. [9] have all shown that in autoimmune thyroiditis, antibodies are found which will localize in the colloid and secretory epithelium. This is true both in man and, in an experimental situation, in the rabbit. White has also ingeniously demonstrated antigen and antibody simultaneously in cases of autoimmune thyroiditis [94a].

Marshall reported briefly on pancreatic deoxyribonuclease and ribonuclease. The antisera were complex and gave both an apical and general localization [57]. Lacy [50b] has shown that insulin is synthesised in the cells of the pancreas. The immunohistological data are supplemented by E. M. data.

Serum proteins

The distribution of serum proteins native to the tissue has been examined by Gitlin, et al. [37] who made fluorescent antisera to human gamma globulin, albumin, metal-combining globulin, fibrinogen and beta lipo-protein with which they treated sections of several organs. With the exception of fibrinogen which was confined to the circulation, these proteins were found in a bewildering variety of tissues and most frequently in the nuclei. If we examine their results in liver and kidney alone we find that in liver some nuclei contain metal-combining protein in central spots, others contain lipo-protein diffusely in the nuclei, and round the nucleolus, and a few nuclei contain gamma globulin and fibrinogen. The cytoplasm of liver cells may contain gamma globulin or albumin. The kidney contains lipo-protein in some nuclei in the glomerulus and tubules, and a small number of nuclei in glomerulus and tubules contain gamma globulin, albumin or metal-combining globulin. Evidence that the synthesis of many serum proteins takes place in the liver or of gamma globulin in the spleen and lymph nodes would probably be accepted without question. The wide nuclear distribution of different proteins (and that of many injected foreign proteins, [reviews 20, 58, 76] indicates either that many nuclei must take up and accumulate serum proteins readily from outside, in which case we must assume that the passage of these proteins through the cytoplasm is so rapid, or the quantities in passage are so small that cytoplasmic detection has not been possible, or that there are actually strong cross reactions between certain serum proteins and some nuclear constituents. That this is the case has been suggested by Schechtman and Nishihara [77]. Schiller *et al.*, however, found rat serum proteins in the cytoplasm of rat cells and not in the nuclei [79]. (The major part of their paper is concerned with the fate of foreign protein). Nairn *et al.* found no serum proteins in any non-pathological liver cells in the rabbit except for the cytoplasm of Küppfer cells. Their fixation procedures were different from those used in other serum protein studies [64]. Whatever the cause, there appears to be a difference between murine and rabbit nuclei and human ones in serum protein content.

The distribution and synthesis of gamma globulin present a case of special interest; (Coons *et al.*, White *et al.* [21, 54, 93, 94]), firstly because there is little doubt that by using the sandwich method only the antibody gamma globulin is being detected; secondly because synthesis can be stimulated by antigenic injections and different stages in the development of synthetic processes in the cell readily examined. The earliest stage in the development of a mature plasma cell is a cell with a faint fluorescent rim of cytoplasm around a large nucleus. The volume and brightness of the cytoplasm increases until in the mature plasma cell the cytoplasm is bright except for the golgi apparatus. Fluorescent bodies are frequently seen in the nuclei, mainly in immature cells, and take the form of rods, dots and strands (Fig. 4). The frequent existence of groups of cells whose central members are brighter than those at the periphery suggests that the progeny of an antibody-forming cell may themselves also become antibody secreters. In some rare cells in the lymph nodes, the plasma membrane, the nuclear membrane, cytoplasmic granules, and rod-like objects in the nucleus all appear bright. These data raise the question of the synthetic significance of the cellular localizations: for example are the bright nuclear structures containing globulin alcohol fixed chromosomes or nucleolus-associated material? Isolated chromosome preparations might be of interest here. Coons discusses the sequence of events in

[23a]. However, Ortega and Mellors [67] find that the same sets of cells contain nonimmune gamma globulin; but that nuclei are uniformly negative. This discrepancy might be technical, but is more probably due to the recent antigenic stimulus to active synthesis in the first case. Another important question is whether in an animal injected simultaneously with several antigens one cell can make more than one type of antibody, or whether it is restricted to making one at a time. This may be a problem of synthetic ability or a problem in cellular acceptance of the antigenic stimulus; possibly one successful stimulation

Fig. 4. Antibody formation.

makes a cell refractory to further stimulation. Nossal and Lederberg [65] and Nossal [66] have approached this problem by trying to estimate the antibody production of single cells in hanging drops. The experiment demonstrates clearly that the antibody production (against Salmonella strains) of single cells may be measured, but is so scored (positive—100% immobilization; negative, any other result) that it is not possible to decide with certainty whether the cells are indeed able to synthesize only one type of antibody molecule or whether a few cells can synthesize two but in very small quantities of each. White has recently convincingly demonstrated [90] and [91] by a technique in which two antigens were used as a "sandwich" between the antibody-containing cells and the two antisera to these antigens, each of which was labelled with a contrasting fluorochrome, that separate cells manufacture antibody to ovalbumin and diphtheria toxin respectively. It remains to be seen whether a cell can make several antibodies to closely related antigens; since separate antibodies can be made to a protein with two haptens it seems possible that in some circumstances it could do so. *

Some other findings on native serum proteins must be mentioned. Fibrin has been demonstrated in scar tissue [36] and some properties of rheumatic sera [2] and of lupus erythematosis sera described [1, 44, 33, 34, 35]. The sera of lupus patients has been shown to contain antibodies to nucleo-proteins and to localize in the nuclei of cells of normal individuals. Since lupus sera cross-react with bovine nucleo-protein it might be worth determining whether the nucleic acid or the protein moiety has an undue share of the responsibility for this. (Other recent papers on serum proteins include [2a, 66a, and 83a]).

* Note added in proof: See Attardi *et al. Bact. Rev.* **23**, 213 (1959).

Virus antigens

A great part of our understanding if the rôles of nucleic acids and proteins in protein specificity and reproduction comes from studies on virus infectivity, reproduction and relation to the host cell; so that demonstrations of virus with fluorescent antibody are of particular interest. There are several papers on viruses; some of these are primarily diagnostic, but a number show the intracellular localization and in some cases the course of infection; these are reviewed by Coons [23]. Since this important review, a few further cases have been described: Prince and Ginsberg [72] find Newcastle virus disease first on the periphery of cells, and then in the cytoplasm; Mellors [61] describes Shope papilloma virus in nuclei, (but not in the hypertrophied epithelium and Ross and Orlans [74] have made a thorough study of the changes in DNA and virus antigen in herpes infected HeLa cells. Cellular DNA increases 6–9 hours after infection, but they could not detect visible changes until 12 hours, when the host DNA breaks up and diffuse DNA appears around the nuclear membrane; at 16 hours this is associated with virus antigen, which increases and invades the cytoplasm, until a necrotic cell contains large amounts of DNA and antigen. These data confirm Lebrun [53a]. The synthesis of adenovirus has also been followed [6a, 69a].

Muscle

Fink, Holtzer and Marshall [31] applied labelled anti-myosin to glycerol extracted chick muscle and found a brilliant labelling of the A band. The M and H bands when present took up some antiserum. In gel diffusion tests the antiserum showed two minor components in addition to the major anti-myosin components. Klatzo, Horvath and Emmart [49] have repeated the results of Fink *et al.* and have found in addition, that in myotonic dystrophy the striation disappears entirely and the fibre becomes irregular. They examined cardiac muscles also and noted that the intercalated discs remained dark. Kaplan has localized two sarcoplasmic antigens in cardiac muscle; cardiolipin, and another unidentified one. The localization of these two was the same; the nucleus, muscle fibers and inner side of the sarcolemma are outlined [48]. Cruickshank and Hill find that the basement membrane antigen localized in the kidney is also found in the sarcolemma [24]. A composite diagram of the distribution of muscle antigens may be seen in Fig. 5.

Holtzer *et al.* further examined the formation of the myo-fibril [46]. The immunohistological method proved more sensitive than any other for the detection of fine structure in the chick embryonic myofibril. They noted that the fibre when it first appears at stage 15 is close to the sarcolemma and is unstriated. At stage 16–17, the main part of the fibre has striations which are identical with those of adult muscle, but it is not striated at its ends; they think that growth takes place at these ends. Ebert [29] reported that cardiac myosin antisera gave positive precipitins at a much earlier stage in the embryo than that at which Holtzer et al. were able to observe cardiac fluorescence with antisera to skeletal myosin. Although there is some difference in specificity, it is possible that there are some soluble prefibrillar precursors of myosin which are extracted by glycerol. Sections of frozen-dried somites, however, appear the same as glycerol-treated ones. The specificity of the granules described by Moscona would be of interest [62]. The only visualization of muscle antigens in earlier stages with labelled antisera

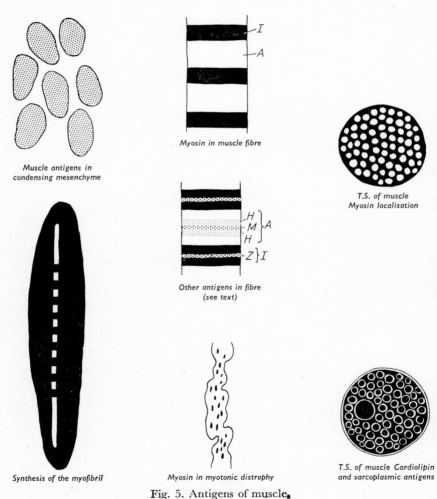

Fig. 5. Antigens of muscle.

was made with an antiserum to whole muscle, which localized in condensing mesenchyme cells in the embryo which certainly do not contain myo-fibers. However, the nature of the antigens which were detected at this stage is not known [13]. Laufer reports briefly on actomyosin in limb regenerates [52]. More recently Holtzer et al. [46a] have found myo fibrils in 3–4 day cardiac myoblasts.

The work has recently been extended by Marshall et al. [57a] who have presented beautiful data on the localisation of myosin in the A band, actin in the I band and M line, tropomyosin in the I band. By splitting myosin (into four distinct fragments) they have already obtained evidence which strongly suggests that the α helical (H) fraction is in the centre of the A band abutting on the M line, while one of the extended ones (L1) is at the other, lateral edge of the A band. Holtzer [46b] has shown that extraction procedures cause the shifting of actin.

Lens

Antisera for lens have been labelled with a fluorochrome or iodine 131. Van Doorenmaalen [27a] and Clayton [12] prepared the antigens by slightly different procedures and the embryonic material was different. Clayton, and Clayton and Feldman [13] worked with the albino mouse and van Doorenmaalen with the chick. In spite of these differences some general features emerge together with some points of specific interest. In very young stages of the chick up to 4 days and the mouse up to about 7 days these antisera stain the entire head structures, the lens vesicle, eye cup, epidermis and brain. Clayton noted in addition that the fluorescence in the eye-cup and vesicle was stronger than that of the brain, and the maximum intensity, still rather weak, was round the inner side of the lens groove and the contiguous head epithelium, and that iodine 131 labelled antisera gave a generally similar picture. Van Doorenmaalen noted that the histological condition of the chick embryo was poor at this stage; the mouse embryos, however, were histologically quite well preserved. Van Doorenmaalen discussed several possible explanations of this weak general staining. Since the antisera were complex, the significance of an antigen of unknown constitution apparently found in a relatively high concentration in adult lens and in a very low concentration in brain and other structures external to the lens in the embryo, remains to be elucidated. It is perhaps parellel to the situation found by Ebert for cardiac myosin [26]. Here also, an antigen was found by precipitin tests to be generally distributed at a low concentration in an early embryo, and to become progressively restricted in localization as the concentration increased during histogenesis. The phenomenon may turn out to be widespread. Some sort of segregation of antigens may underlie the gradual differentiation of tissues and their progressive loss of alternative developmental pathways. As some antigens are (almost certainly) lost, others are certainly synthesized in an increasing amount, and the synthesis of new organ specific ones becomes possible. Thus an adult organ may contain not only organ specific antigens which began to develop after the rudiments had been demarcated, but some which at an earlier stage are found in many tissues, most of which do not contain them in the adult. Some antigens will doubtless remain generally distributed

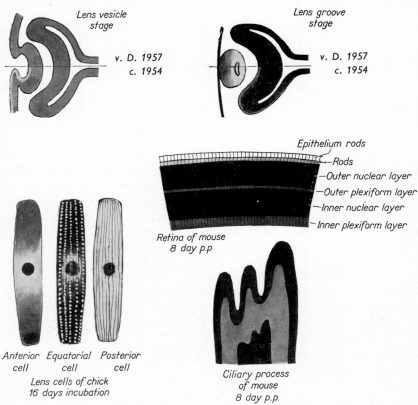

Fig. 6. Lens antigens.

throughout development or even achieve a wide distribution as the result of exchanges and intercellular transport.

Information on the intermediate stages of lens formation are drawn from van Doorenmaalen's data since an equivalent stage in the mouse was not examined. He finds that at five days the staining outside the lens disappears. The lumen fluoresces and at the same time the cytoplasm round the nucleus of the lens cells begins to fluoresce. By five to six days the cells are filled with rows of fluorescent granules, while in the cytoplasm the perinuclear fluorescence increases. Lens fibres may be seen as unstained structures between rows of stained granules. Finally as the fibres increase in size and number the granules disappear and the fibres become evenly stained. He suggests that the antigen localized is the ground substance or fibrillar cement; and in fact the fibres are probably eliminated by centrifugation in his preparation of the original antigen for immunization. The maximum fluorescence of a sixteen day chick lens is round the cavity of the lens vesicle. The general pattern of staining is parallel to the distribution of basophilic granules, and in the anterior region where the fibre formation is less advanced, the granular localization of antigen persists, while in the posterior region the fibers appear uniformly stained. Van Doorenmaalen's sera reacted with iris *in vitro*; however, in immunohistological localization it was negative, except for a bright line on the inner margin of the retina which was sometimes noticed. He considers this to be due to the adherence of lens fibers to the retinal surface. A similar appearance was noticed in mouse preparations of the earlier stages but there was no obvious lens material attached. As in the chick, the maximum fluorescence surrounds the cavity of the mouse lens vesicle. In the newborn mouse the lens fibres stain evenly and very brightly but in addition the retina shows some staining. In preparations labelled with iodine 131 the label was in the fibrillar layers; the maximum intensity being in the outer pigmented epithelium. No sections treated with iodine 131 included the ciliary process, which was treated with fluorescent antiserum. The ciliary pigmented epithelium cross reacted with lens antiserum and the mesenchyme was negative. In the retina the pigmented epithelium was fluorescent and the remaining fibrillar layers weakly so; the nuclei were negative. Subsequent tests suggest that only the labelling of the outer epithelium is significant. The antigen responsible for the reaction of regions of the retina and ciliary process with lens antisera are not known. These antisera were undoubtedly complex; however, Wolffian regeneration of the lens in fact takes place from these regions of the iris [81]. In an *in vitro* study, Langman *et al.* [51a] have investigated lens antiserum crossreactions. Nace and Clarke [63a] report that antisera to supernatant of extract of tail-sub stage stains the eye-cup and lens placode.

Embryonic induction

Whether macromolecules are transferred from the inductor to the induced tissue in organizer action is a problem that has been approached by many different methods (Review, [7]). Attempts to determine this by immunohistological methods using an organizer graft of one species and ectoderm of another species were entirely inconclusive [84]. The species used were somewhat closely related — *Triturus alpestris* in *T. cristatus*, and chick in duck; it is, therefore, possible that in the absorption of the antisera to produce some species specificity, the antibodies to the relevant antigens had been eliminated. An attempt [14] to determine this with guinea-pig liver induction in amphibian ectoderm has indicated that guinea-pig antigens may be found coating the yolk platelets in a few cells, especially if near the inductor. Most, although not all, cells containing guinea-pig antigens were moribund, and since the specimens examined to date were too early to show induced structures the relevance of their uptake to induction is not established, nor has the possibility been excluded that the dying cells liberated some amphibian organizer substance. A few embryos with palisading were evenly stained. These data confirm, in a general way, that of Vainio [96] and Rounds and Flickinger [75] and [31a].

The site of action of cytotoxic antisera

The glomerular localization of nephrotoxic antisera has been described above. White has demonstrated antibodies to thyroglobulin in cells infiltrating the thyroid and in lymph nodes adjacent to it [92]. Hiramoto *et al.* confirm this [43].

Many attempts have been made to culture embryos in antisera; these have been reviewed by Nace [63]. The effects obtained range from negative to general cytotoxicity or even specific inhibitions and the same applies to the effects obtained by injecting antisera *in vivo*. Here, however, physiological factors which may prevent the antibody from reaching the tissue may be involved. In an embryo cultured in antisera the range of results must be due to several variables. The site of the antigen, the permeability or otherwise of the cell [76] (some cells in tissue culture have been shown to pick up and retain foreign specificity [51] and to accept proteins entire probably by pinocytosis [32]). An attempt was made [16] to find the site of action of antisera which had been found to exert an effect on amphibian embryo cultures by cultivating embryos in rabbit antisera or normal sera and treating frozen dried sections with a fluorescent sheep anti-rabbit globulin. Many cells, particularly in the ectoderm, appeared to contain the rabbit specificity; the nuclei were in all cases negative. Frequently the cell membrane was obviously damaged, but it was not determined whether this occurred before or after the entry of the sera. Holtzer [46b] has investigated the problem and found no evidence that antisera with complement destroyed are cytotoxic or can enter any except damaged cells (embryo or adult) except those which normally pinocytose. Weiler [89a] found that his antisera are cytotoxic only in the presence of complement. Some of the discrepancies in the literature are therefore likely to depend on the presence or incomplete removal of complement, or whether the antiserum contains antibodies damaging to the cell membrane. Pinocytosis has been investigated [46b, 57b] with labelled proteins, and by Brandt who has studied the course of pinocytosis of serum by amoebae [8]. With labelled antiglobulin, Clarke [116] has similarly studied the uptake of protein into the gut of neonatal mammals.

Summary

There are a great many problems connected with the study of differentation which may be studied with advantage by means of labelled antisera.

The antigenic mapping of different tissues, for example, is a necessary descriptive task, while the times and sites of synthesis of particular well-defined components during the ontogeny of a tissue or during the course of events in actively synthesizing the cells can be studied by this technique.

Particularly interesting would be the study of the changing relations between two antigens during the differentation of related tissues with diverging synthetic abilities. Antigenic changes during carcinogensis, induction, regeneration or during prolonged tissue culture may also be examined. Cellular permeability to macromolecules is studied with advantage by this method of approach. Another type of problem would be that of estimating the rate of somatic mutation or crossover, provided that a suitable pair of gene-controlled antigens is employed.

The potentialities of labelled antisera as histological agents are considerable, and aside from known technical snags which may be subject to controls, the main limitations are set by the stability and specifity of the antigens themselves.

Acknowledgments

My grateful thanks are due to those authors who have permitted me to quote from unpublished data; to Professor C. H. Waddington for encouragement; to Mr E. D. Roberts and M   for the text figures; to Dr. G. A. Clayton and Miss A. Gr   Dutch; and to the Library and Office Staff   Genetics. My thanks are also due to the pub   the addition of recent references in proof.

References

1. ALEXANDER, W. R. M. (unpublished).
2. ALEXANDER, W. R. M. *Immunology*, **1**, 217 (
2a. BARDAWIL, W. A., TOY, B. C. and HERTIG, (1958).

3. BEALE, G. H. and KACSER, H. *J. Gen. Microbiol.* **17**, 1. 68 (1957).
4. BERENBAUM, M. C. *Nature, Lond.* **177**, 46 (1956).
4a. BERENBAUM, M. C. *Immunol.* **2**, 71. (1959).
4b. BING, J. and WIBERG, B. *Acta Path. Microbiol. Scand.* **44**, 138 (1958).
5. BIRNBAUM, D. and WOLMAN, M. *Experientia* **14**, 6, 204 (1958).
6. BOURSNELL, J. L., COOMBS, R. R. A. and RIZK, V. *Biochem. J.* **55**, 745 (1953).
6a. BOYER, G. S., DENNY, F. W. and GINSBERG, H. S. *J. Exp. Med.* **110**, 827 (1959).
7. BRACHET, J. "Biochemical Cytology." Academic Press, Inc. N.Y. (1957).
8. BRANDT, P. W. *Exp. Cell Res.* **15**, 300 (1958).
9. BUETNER, E. H., WITEBSKY, E., ROSE, N. R. and GERBASI, J. R. *Proc. Soc. Exp. Biol., N.Y.* **97**, 712 (1958).
10. BUSH, V. and HEWITT, R. E. *Am. J. Path.* **28**, 863 (1952).
11. CHADWICK, C. S., MCENTEGART, M. G. and NAIRN, R. C. *Lancet* 412 (1958).
11a. CHADWICK, C. S., MCENTEGART, M. G. and NAIRN, R. C. *Immunol.* **1**, 315 (1958).
11b. CLARKE, S. *J. Biophys., Biochem. Cytol.* **5**, 1. (1959).
12. CLAYTON, R. M. *Nature, Lond.* **174**, 1059 (1954).
13. CLAYTON, R. M. and FELDMAN, M. *Experientia* **11**, 1, 29 (1955).
14. CLAYTON, R. M. and ROMANOVSKY, A. *Exp. Cell Res.* **18**, 410 (1959).
15. CLAYTON, R. M. (unpublished).
16. CLAYTON, R. M. In press.
16a. CLAYTON, R. M. *Exp. Cell Res.* Suppl. 7. 275 (1959).
17. CONN, H. J. "Biological Stains." Biotech. Publ. Geneva, N.Y., U.S.A. (1946).
18. COONS, A. H., CREECH, H. J. and JONES, R. N. and BERLINER, E. *J. Immunol.* **45**, 159 (1942).
19. COONS, A. H. and KAPLAN, M. H. *J. Exp. Med.* **91**, 1. (1950).
20. COONS, A. H. *J. Exp. Biol. Symp.* **6**, 166 (1951).
21. COONS, A. H., LEDUC, E. H. and CONNOLLY, J. M. *J. Exp. Med.* **102**, 49 (1955).
22. COONS, A. H. *Intern. Rev. Cytol.* **5**, 1 (1956).
23. COONS, A. H. in "The Nature of Viruses." Ciba Foundation. Churchill, London. p. 203 (1957).
23a. COONS, A. H. *J. Cell. Comp. Physiol.* **52**, Suppl. 1, 55, (1958).
23b. COONS, A. H. in "General Cytochemical Methods," vol. 1, ed. J. F. Danielli, Academic Press Inc., N. Y. p. 399 (1959).
24. CRUICKSHANK, B. and HILL, A. G. S. *J. Path. Bact.* **66**, 283 (1953).
25. CRUICKSHANK, B. and CURRY, A. R. G. *Immunology* **1**, 1. 13 (1958).
26. CRUICKSHANK, B. *Lancet* **412**, 1. 7017 (1958).
26a. CURTAIN, C. C. *Nature, Lond.* **182**, 1305 (1958).
26b. DINEEN, J. K. and ADA, G. L. *Nature, Lond.* **180**, 1284 (1957).
27. VAN DOORENMAALEN, W. J. Thesis. Amsterdam University (1957).
27a. VAN DOORENMAALEN, W. J. *Acta Morphol. Neederland Scand.*, **11**, 1. (1958).
28. DONIACH, I., and PELC. S. R. *Proc. Roy. Soc. Med.* **42**, 957 (1949).
29. EBERT, J. D. *Proc. Nat. Acad. Sci., Wash.* **39**, 333 (1953).
30. EBERT, J. D. in "Aspects of Synthesis and Order in Growth." Ed: D. Rudnick. Princeton Univ. Press., 69 (1955).
31. FINCK, H., HOLTZER, H. and MARSHALL, J. M. Jr. *J. Biophys. Biochem. Cytol.* **2**, Suppl. 4. 175 (1956).
31a. FLICKINGER, R. A., HATTON, E. and ROUNDS, D. E. *Exp. Cell Res.* 17. 30 (1959).
32. FRANCIS, M. D. and WINNICK, T. *J. Biol. Chem.* **202**. 273 (1953).
33. FRIOU, G. S., FINCH, S. and DETRE, K. *J. Immunol.* **80**. 324 (1958).
   FRIOU, G. S. *Proc. Soc. Exp. Biol. N.Y.* **97**, 738 (1958).
   FRIOU, G. S. *Yale J. Biol. Med.* **31**, 40 (1958).
   ...LIN, D., CRAIG, J. M. and JANEWAY, C. A. *Amer. J. Path.* **33**, 267 (1957).
   ..., D., LANDING, B. H. and WHIPPLE, A. *J. Exp. Med.* **97**, 163 (1953).
   ... E., HOLBOROW, E. J. and JOHNSON, G. D. *Lancet* 7005, 1083 (1957).

38a. GLYNN, L. E. and HOLBOROW, E. J. *Brit. Med. Bull.* **15**, 150 (1959).
38b. GOLDMAN, M. and CARVER, R. K. *Science* **126**, 839 (1957).
39. GOLDWASSER, R. A. and SHEPARD, C. C. *J. Immunol.* **80**, 2, 122 (1958).
40. GOODMAN, M., GREENSPON, S. A. and KRAKOWER, C. A. *J. Immunol.* **75**, 96 (1955).
41. GREENSPON, S. A. and KRAKOWER, C. A. *A.M.A. Arch. Pathol.* **49**, 291 (1950).
42. HILL, A. G. S. and CRUICKSHANK, B. *Brit. J. Exp. Path.* **34**, 27 (1953).
43. HIRAMOTO, R., ENGLE, K. and PRESSMAN, D. *Proc. Soc. Exp. Biol. Med.* **97**, 611 (1958).
43a. HIRAMOTO, R., JURANDOWSKI, J., BERNECKY, J. and PRESSMAN, D. *Proc. Soc. Exp. Biol. N.Y.* **101**, 583 (1959).
43b. HIRAMOTO, R., GOLDSTEIN, D. and PRESSMAN, D. *Cancer Res.* **17**, 1135 (1957).
43c. HIRAMOTO, R. YAGI, Y. and PRESSMAN, D. *Cancer Res.*, **19**, 874 (1959).
44. HOLBOROW, E. J., WEIR, D. M. and JOHNSON, G. D. *Brit. Med. J.* 5047, 732 (1957).
45. HOLBOROW, E. J. (unpublished).
46. HOLTZER, H., MARSHALL, J. M. and FINCK, H. *J. Biophys. Biochem. Cytol.* **3**, 5, 755 (1957).
46a. HOLTZER, H., ABBOT, J. and CAVANAUGH, M. C. *Exp. Cell Res.* **16**, 595 (1959).
46b. HOLTZER, H. *Exp. Cell Res.* Suppl. 7, 234 (1959).
46c. HOLTER, H. and HOLTZER H. *Exp. Cell Res.* 18, 421 (1959).
47. HUGHES, P. E., LOUIS, C. J., DINEEN, J. K. and SPECTOR, W. G. *Nature, Lond.* **180**, 289 (1957).
47a. HUGHES, P. E. *Cancer Res.* **18**, 426 (1958).
47b. JANKOVIC, B. D. and LINCOLN, T. L. *Vox Sang.* **4**, 119 (1959).
48. KAPLAN, M. H. *J. Immunol.* 80, 254 (1958).
48a. KING, E. S. J., HUGHES, P. E., LOUIS, C. J. *Brit. J. Cancer*, **12**, 5 (1958).
49. KLATZO, I., HORVATH, B. and EMMART, W. E. *Proc. Soc. Exp. Biol.* **97**, 1, 135 (1958).
50. KRAKOWER, C. A. and GREENSPON, S. A. *A.M.A. Arch. Pathol.* **51**, 629 (1951).
50a. KRAKOWER, C. A. and GREENSPON, S. A. *A. M. A. Arch. Path.* **66**, 364, (1958).
50b. LACY, P. E. *Exp. Cell Res.* Suppl. **7**, 296 (1959).
51. LANGMAN, J. *Konink. Ned. Akad. Wetenschap.* C. **219**, 214 (1953).
51a. LANGMAN, J., SCHALEKAMP, M. A. D. H., KUYKEN, M. P. A. and VEEN, R. *Acta Morph. Neederl-Scand.* **1**, 142 (1958).
52. LAUFER, H. *Anat. Rec.* **128**, 3, 580 (1957).
53. LEBLOND, C. P. and GROSS, J. *Endocrinology* **43**, 306 (1948).
53a. LEBRUN, J. *Virol.* **2**, 496 (1956).
54. LEDUC, E. H., COONS, A. H. and CONNOLLY, J. M. *J. Exp. Med.* 102, 61 (1955).
55. LOUIS, C. J. *Stain Tech.*, **32**, 6, 279 (1957).
55a. LOUIS, C. J. *Austral. Ann. Med.* **6**, 277 (1957).
55b. LOUIS, C. J. *Austral. Ann. Med.* **6**, 300 (1957).
56. MARSHALL, J. M. Jr. *J. Exp. Med.* **94**, 21 (1951).
57. MARSHALL, J. M. Jr. *Exp. Cell Res.* **6**, 240 (1954).
57a. MARSHALL, J. M., HOLTZER, H., FINCK, H. and PEPE, F. *Exp. Cell Res.* Suppl. 7, 219 (1959).
57b. MARSHALL, J. M. and HOLTER, H. C. R. Lab. Carlsberg, sér Chin. 29, 7 (1952).
57c. MAYERSBACH, H. *Acta Histochem.* **5**, 351, (1958).
57d. MAYERSBACH, H. *J. Histochem. Cytochem.* **7**, 427 (1959).
58. MCMASTER, P. D. "The Nature and Significance of the Antibody response." N.Y. Columbia Univ. Press (1953).
59. MELLORS, R. C., SPIEGEL, M. and PRESSMAN, D. *Lab. Invest.* **4**, 69 (1955).
60. MELLORS, R. C., ORTEGA, L. C. and HOLMAN, H. R. *J. Exp. Med.* **106**, 2, 191 (1957).
61. MELLORS, R. C. *Fed. Proc.* **17**, 714 (1958).

62. MOSCONA, A. A. *Exp. Cell Res.* **9**, 377 (1955).
63. NACE, G. W. *Ann. N.Y. Acad. Sci.* **60**, 1038 (1955).
63a. NACE, G. W. and CLARKE, W. M. in "Chemical Basis of Development." Ed. McElroy and Glass, p. 546 (1958).
64. NAIRN, R. C., CHADWICK, C. S. and MCENTEGART, *M. G. J. Path. Bact.* **76**, 143 (1958).
64a. NAIRN, R. C., FRASER, K. B. and CHADWICK, C. S. *Brit. J. Exp. Path.* **40**, 2 (1959).
65. NOSSAL, G. J. V. and LEDERBERG, J. *Nature, Lond.* **181**, 1419 (1958).
66. NOSSAL, G. J. V. *Brit. J. Exp. Path.* **34**, 544 (1958).
66a. OHTA, G., COHEN, S., SIGER, E. T., ROSENFIELD, R. and STRAUSS, L. *Proc. Soc. Exp. Biol. N.Y.* **102**, 187 (1959).
67. ORTEGA, L. G. and MELLORS, R. C. *J. Exp. Med.* **106**, 627 (1957).
68. ORTEGA, L. G. and MELLORS, R. C. *J. Exp. Med.* **104**, 151 (1956).
69. PEARSE, A. G. E. "Histochemistry." Churchill, London (1954).
69a. PEREIRA, H. G., ALLISON, A. C. and BALFOUR B. *Virol.* **7**, 300, (1959).
70. PRESSMAN, D. *Cancer N.Y.* **2**, 697 (1949).
71. PRESSMAN, D., HILL, R. F. and FOOTE, F. W. *Science* **109**, 65 (1949).
72. PRINCE, A. M., GINSBERG, H. S. *J. Exp. Med.* **105**, 177 (1957).
73. REDETZKI, H. M. *Proc. Soc. Exp. Biol. N.Y.* **98**, 1, 520 (1958).
74. ROSS, R. W. and ORLANS, E. *J. Path. Bact.* **72**, 393 (1958).
75. ROUNDS, D. E. and FLICKINGER, R. A. *Anat. Rec.* **128**, 3, 612 (1957).
76. SCHECHTMAN, A. M. *Intern. Rev. Cytol.* **5**, 303 (1956).
77. SCHECHTMAN, A. M. and NISHIHARA, T. *Ann. N.Y. Acad. Sci.* **60**, 7, 1079 (1955).
78. SCOTT, D. G. *Brit. J. Exp. Path.* **38**, 178 (1957).
79. SCHILLER, A. A., SCHAYER, R. W. and HESS, E. L. *J. Gen. Physiol.* **36**, 489 (1953).
80. SILVERSTEIN, A. M. *J. Histochem. Cytochem.* **5**, 94 (1957).
80a. SINGER, S. J. *Nature, Lond.* **183**, 1523 (1959).
81. STONE, L. S. *J. Exp. Zool.* **136**, 1, 75 (1957).
82. TOBIE, J. E. *J. Histochem. Cytochem.* **6**, 271 (1958).
83. TYLER, A. in "Analysis of Development." Ed: Willier, Weis, and Hamburger (1955).
83a. VAZQUEZ, J. J. and DIXON, E. J. *A. M. A. Arch Path.* **66**, 504, (1958).
84. WADDINGTON, C. H., CLAYTON, R. M., and MULHERKAR, L. Unpublished.
85. WATSON, R. F., ROTHBARD, S. and VANAMEE, P. *J. Exp. Med.* **99**, 535 (1954).
86. WEBER, G. *Biochem. J.* **51**, 155 (1952).
87. WEILER, E. *Z. Naturf.* **76**, 327 (1952).
88. WEILER, E. *Z. Naturf.* **116**, 31 (1956).
89. WEILER, E. *Brit. J. Cancer* **10**, 560 (1956).
89a. WEILER, E. *Exp. Cell Res.* Suppl. 7, 244 (1959).
90. WHITE, R. G. Unpublished.
91. WHITE, R. G. *Nature, Lond.*, **182**, 1383, (1958).
92. WHITE, R. G. *Proc. Roy Soc. Med.* **50**, 11, 953 (1957).
93. WHITE, R. G. *Brit. J. Exp. Path.* **35**, 365 (1956).
94. WHITE, R. G., COONS, A. H. and CONNOLLY, J. M. *J. Exp. Med.* **102**, 23 (1955).
94a. WHITE, R. G. *Exp. Cell Res.* Suppl. 7, 263 (1959).
95. WOERDEMAN, M. W. in "Biological Specificity and Growth." Ed: Butler. Princeton Univ. Press, 1955.
96. VAINIO, T. *Ann. Acad. Sci. Fenn.* **A4**, 35, 94 (1957).
97. YAGI, Y. and PRESSMAN, D. *J. Immunol.* **81**, 7 (1958).

A BIOCHEMICAL APPROACH TO CELL MORPHOLOGY

W. S. Vincent

*Department of Anatomy, Upstate Medical Center,
State University of New York, Syracuse, New York*

WHILE cytology is classically a descriptive science, the papers in this symposium emphasize the fact that the modern cytologist has almost a compulsive bent toward physiology. As a result, we find ourselves inevitably drawing physiological conclusions from our strictly structural observations, often to our sorrow. On the other hand, the modern science dealing with function has been pre-empted—at least nearly so—by the discipline which is called biochemistry. Often within this group the conceptual status of a biological structure is either dimly or not at all realized. The net result becomes one of morphological conclusions from functional observation, often to their sorrow. I think that we cytomorphologists have allowed ourselves to become so mesmerized by the beautiful interlocking cycles of cellular activity elucidated by the biochemist, as well as by the flashing lights and moving dials of modern instrumentation, that we have unconsciously felt that any morphological observation must be coloured with a functional meaning.

I would like to point out in this brief discussion just one small example of how a biochemical approach to cytology can be of some use in pointing the way toward an answer to a classical, though yet unanswered, cytological problem. The classical problem which I have attempted to answer, and which I want to present is that of the origin of the nucleolus.

The nucleolus or plasmosome is found in all cell nuclei (with few exceptions), usually as a spherical body; one or two or occasionally many more in number. We know that it is formed in relation to a specific area of a certain chromosome or chromosomes. What we do not know is whether the nucleolus is synthesized by that particular area, or whether it is accumulated by that chromosome region, being actually the product or products of other regions of the chromosomes. (For a more complete discussion of this problem see Vincent, 1955). Simply stated the question becomes: Is the nucleolus the product of a localized genetic structure, or is it the product of many genes?

An approach to this problem by the use of purely biochemical means became possible when I was able to isolate nucleoli in large quantities from starfish oocytes (Vincent, 1952, 1957). This isolation is accomplished by using what is now a common procedure called differential centrifugation. Simply, the oocytes are broken up in sugar solutions, centrifuged at various speeds, and depending upon the weight of the particles found in the broken-up cells, one obtains layers of relatively pure granular fractions at various levels in the centrifuge tube. (See Fig. 1.)

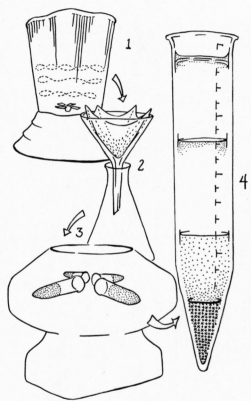

Fig. 1. Schematic figure of the isolation procedure used by the author for the isolation of nucleoli from starfish oocytes.

Mild agitation in sucrose solutions by the blendor (1) serves to break up the oocytes and their fragile germinal vesicle nuclei. Filtration of the homogenate through glass wool (2) removes ovarian stroma and remaining intact cells. Mild centrifugation (3) in the cold (usually at about 200 $\times$ g) is adequate to separate nucleoli from other cell components (4). The nucleoli fraction is usually resuspended and steps 2 and 3 repeated several times until microscopic examination reveals a preparation of satisfactory purity.

A general review of differential centrifugation procedures for cell biologists has been presented by Anderson (1956).

Fortunately, the nucleolus turns out to be the heaviest portion of the cell, thus it is always found on the bottom of the centrifuge tube. It is an easy task, therefore, to prepare very pure preparations of starfish nucleoli which can be analyzed by conventional chemical procedures.

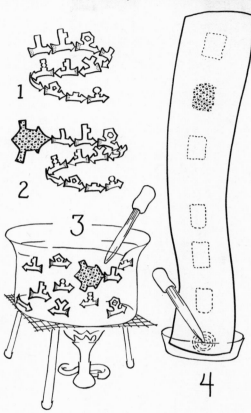

Fig. 2. This figure represents the chain of events in dinitrofluorobenzene (DNFB) labelling of an amino (N) terminal amino acid in polypeptide chain (1). In alkaline aqueous media the DNFB reacts with the free amino group forming a highly stable bond with the liberation of hydrofluoric acid (2). The remaining dinitrophenol (DNP) is shown attached to the amino group (stippled) of the terminal amino acid (arrowlike). Acid hydrolysis at elevated temperature (3) breaks all of the peptide linkages and leaves the DNP label (cross-hatched) attached to the terminal amino acid. By the use of paper chromatography (4) the yellow-coloured DNP-amino acid is separated from the non-coloured subterminal ones, to be identified by comparison with known standards. Note that only the terminal amino acid is shown to carry a free amino group which will react with DNFB. When certain of the diamino-amino acids are found in subterminal positions their free amino group will also react with the DNFB; when found in the terminal position these same amino acids will be doubly labelled. These different categories can be recognized by their different rates of movement on the chromatograph.

Subsequent analyses of these isolated nucleoli revealed that they were mostly protein. It occurred to me that if one could analyze this protein and find out how many types or species of protein it contained, one might get some answer to the question which was asked above. This presumption is based on the notion that a single genetic region

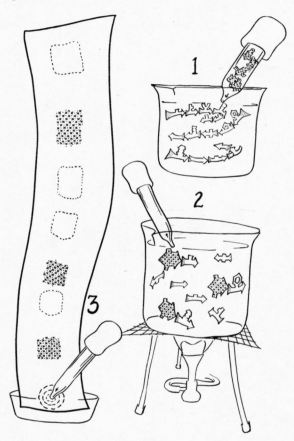

Fig. 3. The labelling of a mixture of protein species is shown in this schematic presentation. At (1) DNFB is added to a mixture of protein molecules whose terminal amino acids are all different. Differing amino acids are indicated by the various lateral attachments to the arrows whose stippled tail represents the amino group. The free DNP-amino acids are shown (2) as they are liberated by hydrolysis. In (3) the separation of the amino acids by paper chromatography reveals the existence of three different terminal groups, thus most likely demonstrating the presence of three differing protein species in the original sample. Note that two different proteins with the same terminal amino acid would not be differentiated, and that a "two-headed" molecule such as insulin would appear as two different molecular species.

An excellent outline of this and other amino acid labelling procedures is given by Cowgill and Pardee (1957).

would most probably produce only a single protein species: and vice versa, materials produced from many different chromosomal areas would most likely be of differing protein species.

We took advantage of the procedure developed here in Great Britain by Professor Sanger in his elegant analysis of the structure of insulin—that of colouring the terminal amino acid of the proteins by reacting them with dinitrofluorobenzene (Sanger & Tuppy, 1951). The reaction will colour the last free amino group of the protein a bright yellow; the protein is then broken up into individual amino acids and the mixture separated by paper chromatography (Fig. 2). Thus if a mixture of proteins is found, one will most likely find a number of yellow spots, each of which represents the terminal amino acid of a particular protein species (Fig. 3). If only a single protein species is present, then only a single yellow spot is found.

When nucleoli from which the ribonucleoprotein complex has been removed (leaving about 85% of the dry weight of the nucleoli intact) are reacted with the reagent, the hydrolysate, upon chromatographic separation, yielded only a single yellow spot. The ribonucleoprotein complex yielded only labelled histidine; isolated yolk platelet protein only labelled glycine; and an acetone powder of whole starfish eggs some half-dozen labelled amino acids.

Although the possibility is not yet completely excluded that the free amino end groups of some protein species may not be available to the dinitrofluorobenzene reagent, the simplest interpretation of these experiments is that most of the nucleolus consists of a single protein. If this simple explanation is true, then it is likely that the nucleolar protein is the product of a single, or closely related genetic loci, rather than being the accumulated products of many different chromosomal regions.

Thus the relatively simple experimental approach described here indicates that biochemical techniques, when applied to a biological question, can yield useful information relating not only to function but to problems of classical morphological cytology.

Acknowledgments

Attendance at the XVth International Congress of Zoology was supported by an American Institute of Biological Sciences travel grant. Part of the work reported was supported by Atomic Energy Commission Grant AT (30-1) to the Marine Biological Laboratory. The drawings were executed by Bertrand Bensam.

REFERENCES

1. ANDERSON, N. G., In "Physical Techniques in Biological Research." (G. Oster and A. W. Pollister, eds.). Vol. 3, pp. 300-352. Academic Press, New York. (1956).
2. COWGILL, R. W., and PARDEE, A. B. "Experiments in Biochemical Research Techniques." pp. 46-61. John Wiley and Sons, New York. (1957).
3. SANGER, F., and TUPPY, H. The amino acid sequence in the phenylalanine chain of insulin. *Biochem. J.* **49**, 463-490. (1951).
4. VINCENT, W. S. The isolation and chemical properties of the nucleoli of starfish oocytes. *Proc. Nat. Acad. Sci., Wash.* **39**, 139-145. (1952).
5. VINCENT, W. S. Structure and chemistry of nucleoli. *Int. Rev. Cytol.* **4**, 269-298. (1955).
6. VINCENT, W. S. Heterogeneity of nuclear ribonucleic acid. *Science* **126**, 306-307. (1957).

PAPER CHROMATOGRAPHY IN RELATION TO GENETICS AND TAXONOMY

A. A. BUZZATI-TRAVERSO

Istituto di Genetica, Università di Pavia, Pavia, Italy

and

Scripps Institution of Oceanography, University of California, La Jolla, California

SUMMARY

PAPER-PARTITION chromatography has been applied successfully to the study of some genetic and taxonomic problems in animals, plants and micro-organisms. A survey is presented of the main results so far obtained in this area of investigation, and some original data pertaining to the subject are submitted. The relationships between this evidence and those resulting from the use of other biochemical and immunological techniques are discussed and analyzed in terms of their significance for genetics, physiology and taxonomy. The potentialities of the paper-chromatographic method are discussed and an outline of further investigations aiming at the solution of some genetic and taxonomic problems is presented.

I. INTRODUCTION

Because of its simplicity, versatility and cheapness, the technique of paper partition chromatography has been effectively used in many kinds of biological investigation since the publication of "Qualitative analysis of proteins: a partition chromatographic method using paper" by R. Consden, A. H. Gordon and A. J. P
the course of the last few years this technique
fully to the analysis of some genetic and
what follows is an attempt to present a surv

The technique of paper partition chromat
of a few simple operations. A small drop of t
substances to be separated is applied to a st
distance from one end, and allowed to d
filter paper nearest to the dried spot is th

oping solution. By capillary action the solvent will flow along the strip and down through its length. In so doing it will remove molecules of the different substances contained in the spot at speeds which are typical for each particular substance and each particular solvent. After a certain time and after the front of the solvent has travelled down the paper, each substance will have moved a characteristic distance, proportional to its speed of movement, from its original position. The relative speeds of the various substances can be measured by calculating the ratios between the distances moved by each spot and the distance covered by the front of the solvent, measured also from the original position. This ratio is called the R_F value of that substance. In this way, for any one solvent each separated substance can be distinguished by its R_F value and by its peculiar chemical properties. Thus, if the original spot was composed of a mixture of amino-acids, e.g., arginine, α-alanine, and leucine, and if a solution of butanol and acetic acid was allowed to flow in the filter paper strip, and after removal from the jar where the partition took place and drying of the paper, then one can identify the three substances by: (1) visualizing them by spraying the paper with a ninhydrin solution which by reacting with the amino-acids will produce coloured spots on the paper, and (2) by measuring the R_F value which will be about 0.19 for arginine, 0.39 for α-alanine and 0.72 for leucine. If one applies one spot of the substance to be analyzed near one corner of a square paper and allows one solvent to flow first in one direction, and then another solvent to flow at right-angles better resolution than with one-dimensional runs can be obtained. With this technique, very small quantities of chemical compounds can be detected, of the order of a microgram or less, and one can thus analyze very small amounts of mixtures.

Very extensive and detailed descriptions of this technique and its modifications, and of the various sets of solvents and developing substances which can be used for the analysis of many organic and inorganic molecules are given in the many laboratory manuals now available [6, 7, 9, 14, 20, 110]. I will therefore refer the reader to these for more technical data.

I wish to stress, however, that this method represents a wonderful tool for biologists who have not available the complicated apparatus that characterizes a modern biochemical laboratory. Only the following were needed to obtain many of the very interesting results to be discussed here: samples of the organisms to be investigated, a few solvents, some filter paper, a few jars or simple cabinets, a drying oven and an ultraviolet lamp.

In the study of genetic and taxonomic problems the technique was simplified by directly applying onto the filter paper without previous chemical manipulations, either whole organisms such as *Drosophila* or parts of them [62] or bits of tissues of larger animals and plants [12]. After these are squashed on the paper and dried they are subjected to partition chromatography. Care should be taken to use solvents appropriate for the separation of the substances present, avoiding if possible the interference of salts and proteins and other disturbing products. This simple and crude technique has given remarkably interesting results, notwithstanding the horror manifested at times by professional biochemists. The relative merits of more or less elaborate techniques of analysis will be discussed in the concluding section of this paper.

II. Paper Chromatography in Genetic Research

Paper partition chromatography has become so widely used in biochemical research that by now there are few experiments in which the biochemist does not use it. This holds true also for biochemical genetics: accordingly there are already innumerable papers in this field in which the author mentions using this technique for the identification of previously known chemical compounds. No attempt will be made here to review such instances. The present survey will focus attention only on those contributions in which the technique of paper chromatography has been used as a tool in genetic research, i.e., either to reveal new aspects of the action of the genetic material or to elucidate previously less well understood physiological and biochemical processes controlled by the genotype.

1. *Drosophila melanogaster*

The paper by Hadorn and Mitchell [62] on "Properties of mutants of *Drosophila melanogaster* and changes during development as revealed by paper chromatography" opened this field of investigation. By conducting paper chromatographic analysis of eggs, larvae, pupae and adult individuals or of their various body sections and organs of *Drosophila*, these two authors discovered a number of chemically unidentified substances having constant and typical R_F values. These could be seen under ultraviolet light by their fluorescence or as coloured spots when the chromatograms were developed with ninhydrin (triketohydrindene hydrate) and were found to vary in number or amount in different tissues at different ages and in different genotypes. These substances, to which preliminary conventional names were given, were later studied chemically in greater detail and in some instances their structure was determined.

The main results obtained by Hadorn and Mitchell can be summarized as follows. (1) During development from the egg to the adult fly, major changes occur in the type and amount of ninhydrin-positive substances; (2) Fluorescing substances change also remarkably during development, as shown in Fig. 1 taken from the paper under review; (3) Very clear-cut differences are apparent in the fluorescent patterns of males and females; (4) Differences in both groups of substances are found regularly in different organs; (5) Mutants affecting the eye

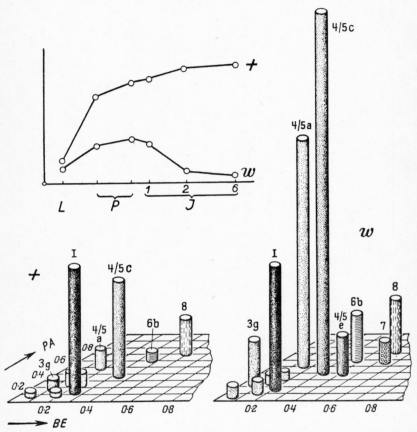

Fig. 1. Biochemical differences between normal genotypes (+) and the white mutant (w) of *Drosophila melanogaster*. The curves represent the mean quantities of isoxanthopterine present in males of different developmental stage; L, larvae before metamorphosis; P, younger and older pupae; J, 1, 2, 6 imagoes one, two and six days after hatching. The diagrams with the columns show the set of fluorescent substances found in corresponding quantities of meconial excrements, for + (left) and w (right). Distribution with R_F values in two-dimensional chromatograms developed in n-propanol-ammonia (PA) and butanol-acetic acid (BE). The heights of the columns indicate the relative quantities as measured fluorometrically (from Hadorn, 57).

colour of the adult fly differ markedly in fluorescent substances; (6) No difference was found in the fluorescent patterns of mutants having different body colours; (7) The chromatograms of larvae of the lethal genotype *ltr* (lethal-*translucida*) showed the presence of many more amino-acids and peptides than the normal control.

These observations were later expanded by various authors. Thus, while the writer [12] confirmed the differences in the fluorescent chromatograms of the two sexes, Fox [44] found that normal males from an isogenic strain possess seven ultraviolet-fluorescent or absorbing substances not observed in normal females; while these possess two fluorescent or absorbing substances not observed in males, ten additional such substances are common to both sexes. Fox showed also that while the two sexes also possess 19 amino-acids, and four unidentified ninhydrin-positive substances, normal males also possess a peptide, referred to as the "sex-peptide" which is not observed in normal females. According to this author the chromatographic differences between males and females cannot be attributed to the Y-chromosome. Very recently Kaplan, Holden and Hochman [75] demonstrated the occurrence of about twice as much free methionine in ethanol extracts of *Drosophila* females than in males.

Expanding their own investigations E. Hadorn and H. K. Mitchell [53, 57] as well as other research workers collected in recent years a large number of data which well substantiate the other observations summarized above. Some of the fluorescing substances have been identified as pteridines [28, 30, 31, 39, 40, 41, 42, 54, 55, 91, 111, 112, 113, 114, 115, 116, 117]. A quantitative study of these substances made possible a better understanding of the formation of the eye pigments of *Drosophila*. These are composed of two groups: the so-called brown and red pigments. While the origin and the first steps in the biosynthesis of the former group were already fairly well understood before the introduction of paper chromatography only this technique has made it possible recently to distinguish and identify the action of mutants which block the last steps in the synthesis of ommochromes. These mutants could not be distinguished by means of feeding experiments and transplantation tests. The red pigment complex is a complicated mixture, whose components are difficult to separate and chemically unstable. For this reason they could not be investigated by ordinary chemical procedures. As late as 1951 no clues were available for understanding the chemical constitution of the red pigments.

When the problem was approached in a different manner, making use of paper chromatography, rapid progress occurred; it was shown that the mutant *sepia* contained a much greater quantity of a yellow pigment than the wild type. Since this mutant does not produce any of the red pigments, the assumption was made that this yellow pigment might be an intermediate in their synthesis or might be closely related to them. Forrest and Mitchell [39, 40, 41] succeeded in obtaining the yellow pigment in crystalline form and in sufficient quantity for further chemical study. In this way they reached the conclusion that the isolated substance was a pteridine, called pteridine-s having the following structure:

$$\text{structure with CO.CHOH.CH}_3, NH_2, CH_2, COOH, OH \text{ groups on pteridine ring}$$

Viscontini, Karrer, Hadorn and others have subsequently isolated several fluorescent substances from *Drosophila* and identified them as pteridines [111]. Three components of the red pigment were also isolated [113] and identified, thus proving the occurrence of a pteridin nucleus in their molecules, which by oxidation or after alkaline hydrolysis gave a common product, 6-carboxy-pteridinic acid. They are: drosopteridine, neodrosopteridine and isodrosopteridine.

By a simple procedure of two-dimensional chromatography the behaviour of the components of the red pigment in mutants affecting eye colour of *Drosophila* has recently been studied in this laboratory (Baglioni, unpublished results). Thus it was shown that there were two more pteridines, which possessed chromatographic and electrophoretic characteristics and ultraviolet spectra similar to those of drosopterine and isodrosopterine, but which could be distinguished from these two substances by their different R_F values using a 20% KCl solution as solvent. The three major components isolated by Viscontini were found to be present in constant ratios also in those mutants which possess an amount of red pigment much smaller than the wild type, while the two new pteridines isolated by Baglioni behave independently from the other components of this group.

On the basis of the extensive investigations of Hadorn and his group on several mutants affecting eye colour the conclusion was reached that each mutation influences the number of different substances in the organism in a specific way, thus determining a pleiotropic action of the gene at the biochemical level [54, 60, 61, 62, 63, 64, 66].

Striking variations in the occurrence of certain fluorescent substances were also observed in the course of development. Fig. 1 shows how the amount of isoxanthopterine varies in wild type flies and in white-eyed ones, and the difference between the corresponding fluorescent patterns in the excretion products of the two genotypes, which is revealed by chromatographic analysis of the meconium deposited by the flies after emerging from the pupa [61]. A paper by Danneel and Zimmermann [24] on the distribution of fluorescent substances in the course of development of several eye colour mutant strains brings further evidence on the pleiotropic effects at the biochemical level of the genes involved. Further detailed data on differences in the concentration of several pteridines measured fluorometrically in the wild type, the *sepia* and *white* mutants and their combinations, showed that the chemical effects of the factors *se* and *w* are of a genuine locus-specific nature [23, less 122].

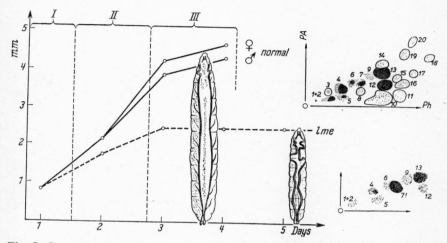

Fig. 2. Growth curves of normal and lethal meander larvae (*lme*) and content of amino-acids, peptides and amides in body extracts of the two genotypes, as revealed in two-dimensional paper chromatograms. Above, the normal; below, the *lme* pattern. The solvents used were n-propanol-ammonia (PA). 1+2, 5, 8: peptides, 3: aspartic acid, 4: glutamic acid, 6: serine, 7: glycine, 9: threonine, 10: lysine, 11: arginine, 12: glutamine, 13 and 15: α- and β- alanine, 14: tyrosine, 16: histidine, 17: γ-amino butyric acid, 18: proline, 19: valine, 20: leucine and isoleucine. The substances missing in *lme* but present in normals are encircled. Note the accumulation of glycine (7!) in *lme*, (from Hadorn, 57).

The work on the effects of lethal genes was also extended. Not only was the case of the lethal-*translucida* analyzed further [16, 65, 109] but also the lethal-*meander* mutant (*lme*) provided more evidence for the interference in the protein metabolism produced by these lethal genes [18, 56]. Fig. 2 illustrates this latter case.

As soon as the first results of their investigations were announced by Hadorn and Mitchell, it appeared of interest to check whether this technique could be used for the investigation of other genetic problems.

Thus the writer [11, 12] attempted to find out whether different genotypes could be distinguished at the paper chromatographic level when their phenotypes were indistinguishable or when mutant phenotypes differed from the wild type in some morphological trait, i.e., when the end product of the mutated gene by which it can be recognized was of a less obvious chemical nature than in the series of mutants studied by Hadorn and Mitchell which affected eye or body colours.

The main results of my investigations can be summarized as follows. (1) Extensive tests on the constancy of results were carried out on a large number of chromatograms obtained from single flies belonging to many different strains, and proved that while, during the first days of adult life, the fluorescent patterns undergo appreciable change, after the tenth day they become constant. Feeding experiments with different strains of yeasts, with moulds and also a comparison between flies grown on a medium containing live yeast and on a completely synthetic medium (unpublished results of the author) proved that the biochemical patterns obtained by this procedure are highly independent of dietary and environmental conditions. (2) Comparison between four unrelated wild type strains showed that these possessed appreciably different fluorescent patterns under ultraviolet, but were identical when the ninhydrin-positive substances were compared. (3) Comparison of the ninhydrin-positive patterns obtained from various *Drosophila melanogaster* strains did not show any appreciable difference. (4) Comparison of a series of several dozen chromatographic patterns obtained from single individuals of a laboratory strain (Oregon-R) and of a highly inbred line derived from it revealed that in the latter the individual variability was much smaller than in the former and that, correspondingly, the inbred strain has much more sharply clear-cut spots than the mass cultured strain. (5) Each of 15 mutant strains gave distinctive fluorescent patterns, except for the striking difference in the fluorescent spots which distinguishes the sexes, a remarkable parallelism of results is to be found when comparing the patterns of females and males of the same mutant strain. (6) Analysis of the chromatographic patterns of special strains in which the *forked* and the *giant* mutations had been introduced into a common genetic background allowed us to conclude that the biochemical specifity as revealed by this procedure falls under the control of a single genetic locus. (7) Individuals heterozygous for completely recessive genes gave, in every case, a chromatographic pattern which differed from that of their wild homozygous parent as well as from that of the parental homozygous recessive. (8) In three cases, maternal effects were evident in the chromatograms of hybrids between wild and mutant strains.

These results, which were obtained while working at the University of California, Berkeley and La Jolla, were later confirmed by further unpublished work carried out in this laboratory. Other authors also obtained similar results. Thus Fox [43, 46] found that there are no appreciable differences in the ninhydrin-positive patterns obtained from one wild type and a dozen strains containing various combinations of the *lozenge* pseudoalleles. Goldschmidt [48] observed differences in the fluorescent patterns of two wild type laboratory strains and the *eyeless*^{2} mutant. By repeated backcrossing to the two wild type strains this author showed that after five generations the fluorescent pattern of the eyeless flies could not be distinguished from that of the normal strains with which they shared a common genetic background. These data, however, do not seem at variance with the conclusions reported above concerning the biochemical specificity determined by singly gene differences. The fact that different wild type strains possess different chromatographic patterns indicate clearly that their biochemical phenotype, as revealed by this technique, is controlled by many genes, and this is also indicated by the differences found between individuals derived from highly inbred and mass cultured strains. The overriding effect of the *forked* and *giant* mutants over the common genetic background which we have just discussed may not always occur. The facts observed by Goldschmidt could then be ascribed to the overriding action of the genetic background of the two chosen strains over the *eyeless*^{2} mutant. It should also be noted that while this author studied the chromatograms obtained from adult heads, mine were obtained from the flies abdomens and thoraxes.

During the larval stage, Koref and Brncic [79] found clear and constant biochemical differences as revealed by ninhydrin-positive substances between individuals belonging to two different strains each of them homozygous for two recessive genes. Larvae with tumours did not differ in their chromatographic pattern from that of their parental strains.

Differences between wild type strains were recently confirmed by Robertson and Forrest [99] who compared the content of isoxanthopterin in mass cultured strains in inbred lines and in their crosses. They came to the conclusion that the amount of this substance is controlled by many genes.

By an analysis of the chromatograms of eggs of *Drosophila* obtained from reciprocal crosses between wild type and *vermillion* mutant flies Graf [49] showed that the amount of kinurenin present is determined by the genotype of the mother and is independent of the genotype of the egg.

Hadorn [58] has determined the amount of drosopterines present in transplanted and host eyes of *Drosophila*, and has found that a wild type pupa is self sufficient for the synthesis of the red eye pigments required for at least 10 eyes.

Hoenigsberg and Castiglioni [71, 72, 73] report that the chromatographic fluorescent patterns of inbred flies show a greater variability during the ageing process of the adult than mass cultured flies. The results appear at variance with those reported by the writer (see point 4, page 102). On the fact that the comparison was made by these workers between mass cultured strains with the *cardinal* mutant and inbred strains with wild type eyes, does not seem to warrant their conclusion, since the *cardinal* mutant is known to affect drastically the metabolism of tryptophan.

The identification of heterozygotes for recessive genes by means of chromatography has been attempted by Kikkawa [76] but he failed to obtain results consistent with those I had previously reported. Further investigations on this point carried out in this laboratory using both uni-dimensional and two-dimensional chromatograms confirmed my original data. A very recent paper by Ziegler-Günder [122] has further evidence to show that, indeed, recessive genes can be identified when in heterozygous conditions, by the effect they exert on the biochemical make up of the organism, even though no difference appears at the phenotypic level.

2. *Other insects*

The interest of the results obtained on *Drosophila melanogaster* prompted research on other species. While more detailed data about the chromatographic patterns of different species of the genus *Drosophila* will be reported in section III of this paper, it is worthwhile to mention here that all species studied by Rasmussen [95, 96] presented clear differences between males and females in the fluorescent substances. While this also proved to be the case for the pupae of *Aedes aegypti* [100] several other insects belonging to *Diptera Culicidae* did not show any sexual dimorphism at the chromatographic level.

In a series of papers on the meal moth *Ephestia kühniella* Hadorn and Kühn and coworkers [37, 59, 60, 80, 81, 82] found that in this species also there is a widespread phenomenon of biochemical pleiotropy, as previously found in *Drosophila*. It was found that a single gene mutation may alter the biochemical constitution of every organ of the individual in a very distinctive way, according to the local biochemical and physiological conditions and structure in different cell systems which build up a body. Fig. 3 illustrates some of the findings on this species.

Laven and Chen [83] working on a lethal mutant of the mosquito *Culex pipiens* discovered facts very similar to those found in the two lethal mutants of the fruit fly, which we have discussed. In this case also, the action of the mutation consists of an interference with the protein metabolism of the larva. The ninhydrin-positive chromatograms

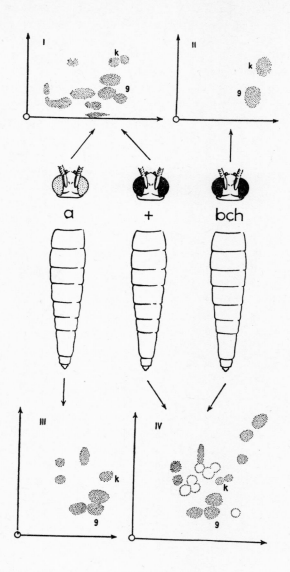

Fig. 3. Organ specificity of biochemical pleiotropy in three genotypes of *Ephestia kühniella*, *a*, + and *bch*. The diagrams above the heads show the set of fluorescent substances present in eyes as revealed by two-dimensional paper chromatograms (I for *a* and +; II for *bch*). Below we see the inventory of substances present in abdomes (III for *a*; IV for + and *bch*). The letters near two fluorescent spots indicate isoxanthopterine (g) and 2-amino-6-oxypterine (k). (from Hadorn, 57).

of methanol extracts of the wild and of the mutant strains show a very marked decrease in the number and amount of amino-acids and peptides in the latter as compared to the former.

A preliminary report by Kikkawa, Ogito and Fujito [77] report that the paper chromatographic analysis of extracts from eggs of wild type and mutant strains of *Bombyx mori* and also of *Drosophila melanogaster* eggs has made possible the identification of metallic complex salts directly connected with the formation of pigments.

3. *Protozoa*

As yet, there are only two short reports on unicellular animals. In both instances acid hydrolysates of cells were subjected to paper chromatography and only ninhydrin-positive patterns were studied. While Butzel and Martin [10] did not find any qualitative difference in the amino-acids content of different stocks, mating types and serotypes of *Paramecium aurelia* (quantitative studies are announced but the results are not yet known), Finley and Williams [38] found clear cut differences in the sexual and asexual forms of the ciliate *Vorticella microstoma*: the former showed 15 bands in the one-dimensional chromatograms and the latter only 10.

4. *Plants*

Very little information is available on plants also, although the technique of paper chromatography has been widely utilized for the identification of certain chemical compounds in some species [e.g., 52, 68, 93] and also in some mutants of *Antirrhinum majus* by Schütte, Langenbeck and Böhme [103]. Buzzati-Traverso has shown [11, 12] that ninhydrin-positive as well as fluorescent patterns obtained from plant tissues, such as root tips, leaves and anthers made a distinction possible between homozygous wild types and heterozygotes for a recessive gene in tomato and in the muskmelon *Cucumis melo* (see Fig. 4).

5. *Man*

The technique of paper chromatography has been recently used for the study of problems of human inheritance and the results are indeed encouraging. Most research workers have focused their attention on the study of amino-acid excretions, because these can be directly assayed by this technique in body fluids such as urine and saliva and because there are large variations between individuals in amino acid excretion.

In the case of cystinuria, a clinical and hereditary condition first recognized over a century ago by the discovery of a renal calculus of unusual composition, the use of paper chromatography has led to a complete reorientation of research. In a paper which can be considered as the first contribution of paper chromatography to genetic investigations, Dent [33, 34] showed how the study of urine by this method could yield precious information. Dent and coworkers [33, 34, 35, 36] showed that cystinuria should be regarded as capable of being broken down into at least two distinct entities: (1) the classical type with a tendency to cystine stone formation and characterized by the excretion in the urine of excessive amounts of cystine, lysine and arginine; (2) a generalized aminoaciduria, corresponding to Lygnac's disease or

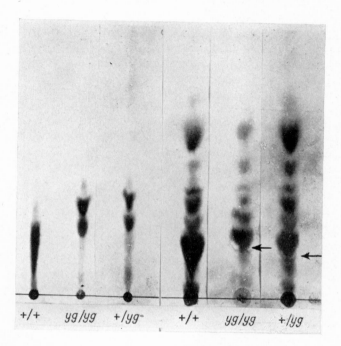

Fig. 4. Chromatographic patterns of ninhydrin-positive materials in muskmelon (*Cucumis melo*) tissues. A. Chromatograms from ground leaves. B. Chromatograms from mashed root tips. Genotypes: $+/+$ = homozygous wild type; $+/yg$ = heterozygous; yg/yg = homozygous for the recessive *yellow-green* mutant. Arrows indicate whitish halo corresponding to very bright blue coloured fluorescing substance present in yg/yg and in the heterozygote (from Buzzati-Traverso, 12).

Fanconi syndrome, with cystine also present as one of the several amino-acids excreted in excess. A third condition in which recessive cystine excretion takes place is Wilson's disease (hepatolenticular degeneration) [35]. The standard tests, for cystine in the urine would permit a discrimination between these distinct pathological conditions, while paper chromatography has made possible the acquisition of this new knowledge.

This technique has been profitably used for the study of other hereditary metabolic disorders. An extensive and clear review of this subject can be found in a paper by Harris [70]. Such investigations have revealed that in addition to pathological conditions, variations in amino-acid excretion occur within the range of normal variation in man. Thus Dent, Harris and others [21, 69] have found a hereditary condition occurring in about 10% of the English population in which large amounts of β-aminoisobutyric acid are present in the urine without producing, however, any abnormal symptom. Studies conducted at the Institute of Human Biology of the University of Michigan [106, 108, 28] indicate that there are racial differences in the occurrence of this biochemical peculiarity in Chinese and Caucasoids, and support the hypothesis of genetic determination of this trait. Other studies on amino-acid excretion in man have been carried out by this technique to discover genetically controlled anomalies within the range of normal variation [3, 4, 5, 47, 118, 119].

It may be of interest for those who are interested in human biology that Rogers [101, 102] has successfully applied the technique of squashing directly small quantities of mammalian tissues onto the filter paper for obtaining chromatograms. These showed constant differences between normal and pathological conditions. Differences were found between different mice tissues, between lymph nodes sampled from rabbits of different age, between lymph nodes of mice with lymphocitic leukaemia and of normal ones, between liver tissues of mice in which fulminating necrosis had developed as a result of the treatment with carbon tetrachloride or of infection with the hepatitis virus of Nelson, in liver tissues of mice made fatty by fasting or by treatment with aminopterin or ethionine, and also in liver tissues poisoned with several toxic agents. The author believes that this technique offers great promise for an understanding of the metabolic background of disease processes, now known for the most part through clinical effects and their gross microscopic appearance and also for discovering chemical compounds effective in blocking or bypassing the primary biochemical injury.

III. Paper Chromatography in Systematics

In 1953 two short notes were published indicating that paper partition chromatography as applied to fresh tissues of several species of fish [13] and in *Drosophila* [12] yielded results that seemed to have value for taxonomic studies. Since that time, a relatively large number of papers have appeared confirming the usefulness of this technique in problems of sytematics and population genetics of a variety of microorganisms and animals, and these will be reviewed now. In Table 1 is presented a list of the studied organisms. No similar attempts appear to have been made in the vegetable kingdom.

1. *Microorganisms*

A number of papers have recently appeared on the results obtained in the study of the taxonomy of lactobacilli [2, 8, 15, 50, 51], streptococci and micrococci, using acid extracts from bacterial cells. Even though the medium and the age of the culture influence the constancy of the chromatograms, this method of analysis seems to offer promise in this group of organisms. While these authors have used extracts from whole cells, Cummins and Harris [22] have analyzed by paper chromatography the hydrolysates of cell-wall preparations of more than 60 strains of corynebacteria, lactobacilli, streptococci, staphylococci and other gram-positive cocci. They found that the amino-acids present in the cell wall seem to be characteristic of the genus, while the sugars and amino sugars seem to characterize the species within the genus. It thus appears that cell wall composition possesses a definite taxonomic value.

Lee, Wahl and Barbu [86] concentrated their studies on the desoxyribonucleic acids of 60 strains of bacteria. They analyzed by paper chromatography the relative amounts of purine and pyrimidine bases present in DNA extracted by first treating with soda, cells which had been ground or lysed by lysozyme and then precipitated after hydrolysis at 175° in concentrated formic acid. The ratio of adenine plus thymine over guanine and cytosine varies through a range between 0.4 and 2.7 and seems to be characteristic of the species. In most species belonging to the same genus in Bergey's classification the ratio is about the same. No relationship was found between the relative amounts of these purine and pyrimidine bases, the Gram character and the aerobic or anaerobic requirements of the bacteria studied.

Different approaches have thus shown that paper chromatography can be of remarkable significance for taxonomic studies in bacteria.

Before leaving this group of organisms it seems worthwhile to mention an early paper which did not have any taxonomic aim but is pertinent to the subject under discussion. While studying the adsorptive behaviour of bacterial viruses Shepard [106] found that mixtures of the various members of the seven bacteriophages of the T series of *Escherichia coli* could be partially separated chromatographically in paper strips which had been treated with hydrochloric acid. The extent of movement up the paper strips of different T-phages has parallelled to a certain extent their other characteristics, such as their serological type. There is a possibility, therefore, that the technique could be extended to the study of other viruses.

Only one study has been made on protozoa. Lee [84] has found that the fluorescent and the ninhydrin-positive patterns of the hydrolysates of *Paramecium aurelia*, *caudatum* and *multimicronucleatum* differ consistently and thus offer a new tool for discriminating between species.

2. *Invertebrates other than insects*

An interesting observation was recently made by Chen and Baltzer [17] working on echinoderms. By paper chromatographic analysis constant differences were found in free- amino-acids and peptides separated from homogenates of eggs from three species of urchins. In spite of their quantitative differences, in all three types of eggs, the chromatographic patterns remained largely constant during development. When, however, the protein precipitates of these species were hydrolyzed in 6N-hydrochloric acid, strikingly similar ninhydrin-positive chromatograms were found. The species-specific differences found in the free amino-acids are not reflected in the amino-acid composition of the proteins. This, according to the authors, may indicate that differences of specific proteins in the sea-urchin egg probably lie on the structural level, as serological studies had previously shown.

In order to test the validity of the procedure described by the writer when applied to an invertebrate phylum, Kirk, Main and Beyer [78] analyzed the fluorescent and absorption patterns under ultraviolet light, of chromatograms obtained from the foot muscle of seven species of Australian landsnails directly squashed on the paper. Chromatography was carried out by using circles of filter paper and feeding the solvent to the centre of the disc. They found clear-cut differences between the species, and these proved to be largely independent of feeding conditions, age and the locality of collection of the animal. These data were later confirmed using an improved chromatographic technique by British investigators [120] who also found this method useful in the study of aquatic snails.

A comparison between the polysaccharides present in aquatic *vs.* landsnails has recently been carried out [91] by chromatography of ethanol extracts of the hydrolyzed tissues. Seven species were studied, and it was found that the galactogen of aquatic forms is different from that of landsnails. In two species, *Lanistes boltenianus* and *Pomacea reteki*, a biochemical sexual dimorphism was found, in that males possessed glycogen while females had galactogen.

3. *Insects*

Since 1951 Micks and coworkers [88, 89, 90] have found constant differences in the free amino acids of mosquitoes. Using one- and two-dimensional chromatography and studying ninhydrin-positive as well as fluorescent patterns, these investigators have been able to show that this technique can be profitable when used not only with mosquitoes but also with three other orders of insects (see Table 1, p. 119). A comparison of density curves obtained from one-dimensional chromatograms proved the high degree of reproducibility of the amino-acid patterns of each species studied. Striking quantative differences were found between chromatograms representing different orders. Appreciable quantitative differences were also found in the comparison of different strains within three species, thus confirming the possibility of using this tool also for population genetic work, as demonstrated in *Drosophila* and fish [12, 13, 98]. Ball and Clark [1] found also similar differences between four species of mosquitoes.

Similarly Robertson [100] on the basis of the results obtained from the study of the fluorescent and ninhydrin-positive patterns of seventeen species of *Coleoptera, Lepidoptera, Diptera* and *Hymenoptera* concludes that this biochemical approach offers a promising means of elucidating problems in taxonomy and population genetics.

An example of such possibilities is given by a recent paper by Merker [87] in which he solves some problems of the taxonomy and distribution of pine mites of the genus *Dreyfusia* using the fluorescent chromatograms obtained by squashing these animals on the filter paper.

Mention should be made also of investigations on the occurrence of pteridines in *Bombyx mori, Colias croceus, Gonopteryx rhamni, Pieris rapae, Vespa crabo, Papilio machao, Opistographis luteolata*. By an association of chromatographic and electrophoretic techniques, interesting results were obtained by Polonovski, Jérome and Gonnard [94] on these insects.

The most complete attempt to evaluate the validity of the paper chromatographic technique for taxonomic studies was carried out by Inge Rasmussen, on the genus *Drosophila*. This seemed to offer an ideal test object because of the very advanced systematic knowledge reached in this group on the basis of morphological, cytological, distributional and genetic data. More than 600 species are known to belong to this genus, and their relative degree of affinity has been established in a very satisfactory way by several students of the subject and especially by A. H. Sturtevant and J. T. Patterson. Within the genus *Drosophila*, subgenera have been established and within some of these a number of "species groups". One can thus find examples within this genus of different levels of taxonomic and probably of phylogenetic relationship. Many living strains of species were available to Miss Rasmussen so that she had ample material available for carrying out such investigation. On the basis of the comparative study of the fluorescent patterns obtained from 19 *Drosophila* species, of *Zaprionus vittiger*, a member of a genus belonging to the *Drosophilidae* family and of *Aphiochaeta xanthina* of the *Phoridae* family, Rasmussen [95, 96] reached the conclusion that

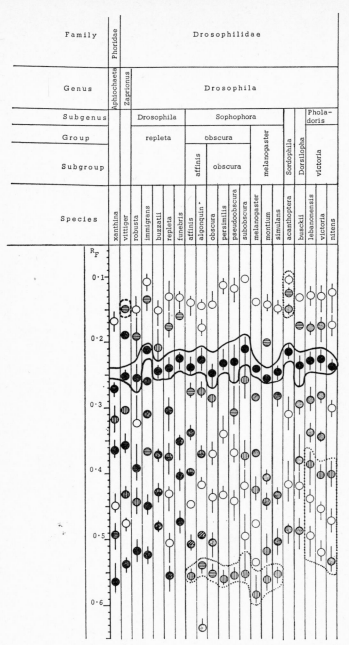

Fig. 5. Schematic representation of the fluorescent patterns obtained in one-dimensional chromatograms of fruit flies. R_F values are given as means of at least 12 readings from a corresponding number of samples of females squashed on the filter paper. Solvent used: 2 parts of n-propanol, 1 part of 1% ammonia. Stripings of the circles indicate different fluorescent colors. (from Rasmussen, 96).

there is a close parallelism between levels of taxonomic affinity as revealed by morphological and other criteria and of biochemical similarity as revealed by these chromatographic studies. Further, chemical compounds and groups of compounds possessing different levels of taxonomic specificity were isolated. Both one-dimensional and two-dimensional chromatographic analyse were used, but no attempt was made to identify the chemical structure of the spots, which have specific or genetic significance. Fig. 5 gives an example of the results obtained. In the only tested case comparable results were obtained using fresh and lyofilized material.

Results comparable to these in insects were also obtained from four species of spiders by Lee, Rene and Roddy [85].

4. *Vertebrates*

Preliminary data on fish [13] showed that: (1) Fluorescent and ninhydrin-positive patterns of the same tissue taken from various specimens of the same species are remarkably constant, irrespective of the size and age of fish; (2) Patterns obtained from muscle of different species show constant and easily recognizable differences; (3) The closer the taxonomic position of the species studied, the greater the similarity of their chromatographic patterns; (4) The same technique can be used to distinguish stocks of the same species belonging to populations geographically separated. These observations were later confirmed by Rechnitzer [98] and Dannevig [25, 26] on the California sardine (*Sardinops coerulea*), 10 *Gadidae*, 3 *Clupeidae* and 4 *Pleuronectidae* and on cod populations. Biochemical investigations by Jones [74] showed also that except for bacterial spoilage, very little change occurs in fish muscle after death, thus opening the possibility of using this technique for field work or market tests.

In 22 mammalian species, Datta and Harris [27] observed differences in the urinary amino-acid pattern. High concentrations of methyl-histidine were found in the urine of all the carnivores: of cystein-S-*iso*pentanol in the urine of several *Felidae;* of cystine in the urine of the Kenya genet (*Genetta tigrina erlangeri*) and of taurine in the mouse, rat, dog, binturong and puma. The observed differences result from complex interactions between dietary, physiological and genetical factors which require a more detailed study.

5. *Interspecific hybrids*

Fluorescent patterns of hybrids between five pairs of species of *Drosophila* and their reciprocal crosses were studied by Scossiroli and Rasmussen [104, 105]. It was found that hybrids show pattern revealing characteristics of both parental species. In some crosses the hybrids show the pattern of the species employed as the mother, but with some minor changes revealing the influence of the father. In some chromatograms of the hybrids a few spots were bigger and brighter than in the two parental species. These spots showed a phenomenon of exaggeration in terms of size and intensity of fluorescence. In three out of these five interspecific crosses, one or more spots not present in either parental species occurred in the chromatogram of the hybrid. It could not be decided whether these "hybrid substances" were already present in the patterns of the parental species at too low concentrations to be detected, or if they were really new and typical of the hybrids only.

Chen and Baltzer [17] report that the ninhydrin-positive chromatograms of the interspecific hybrids between the sea-urchins *Paracentrotus lividus* × *Arbacia lixula* and *Sphaerechinus granularis* × *Paracentrotus lividus* show an entirely maternal character.

IV. Conclusions

The experimental evidence reviewed in the preceding pages indicates without any doubt that the technique of paper partition chromatography is indeed a very useful tool for genetic and taxonomic research. As yet, very few attempts have been made to apply this method of analysis on a wide front and in a systematic way; the majority of the contributions have so far had a cursory character, with the very notable exception of the work of Hadorn and coworkers on the origin of *Drosophila* eye pigments. It appears worthwhile at this stage to extend in depth and in scope genetic and taxonomic research by taking advantage of this chemical tool. A few suggestions for future work will follow.

At the technical level it appears worthwhile to discuss the relative merits of (1) carrying out chromatographic analysis of organisms or tissues directly squashed onto the paper as compared with the study of tissue extracts, and of (2) using one-dimensional or two-dimensional chromatographic analysis. As for the first question, I think it is safe to state that the direct analysis of tissue squashes offers notable advantages in spite of its crudeness. It is simple and can be carried out without any complicated chemical equipment; as a matter of fact it can even be done in the field. It gives surprizingly constant results with a high degree of independence from environmental conditions. It may reveal biochemical relationships that would disappear if one studies extracts or products of hydrolysis of tissues. The case reported by Chen and Baltzer [17] for echinoderms where they found appreciable differences between species in the free amino-acids but not in the products of hydrolysis illustrates this point. Indeed one should keep in mind that in fresh tissues or body fluids there can be substances such as salts or proteins which may interfere with the attainment of clean chromatograms, and proper care should be taken in the choice of the solvents to be used. Sometimes, after a preliminary study of the chromatographic patterns obtained by squashing, it may become necessary to bring the analysis to a higher degree of accuracy and then extracts and hydrolysis products will have to be studied. But this technique can certainly not replace the apparently more crude one. To give only one example, two peptides or protein molecules may possess very different properties in spite of their being made of equal amounts of the same amino-acids. Paper chromatographic analysis of the free peptides as present in the tissues may reveal differences that would disappear if one were to study the hydrolysis product of such peptides.

As for the second question, somewhat similar considerations can be developed. Fox [45] has pointed out the advantages of two-dimensional chromatographic analysis over one-dimensional. The greater resolution obtained, this author maintains, insures better results. One can hardly disagree with this view in general terms, but it appears to me that one-dimensional analysis has its definite merits. For one-dimensional analysis a very small quantity of material is sufficient and in this way one may study single organisms with better results. Sometimes, as in the case reported by Fox, one-dimensional chromatograms fail to reveal differences that appear very evident with two-dimensional analysis, but sometimes things may work the other way around. Like in a kaleidoscope, the mutual interactions between the components of the figure may produce characteristic patterns which would disappear after separation of the parts. Here again, then, both methods of analysis should be used according to the varying requirements of the investigation.

At the genetic level of study a number of problems appear especially interesting for further analysis using this technique. In bacteria it would be very interesting to study how metabolic pathways that have been studied in detail may affect other cell components and how different biochemical conditions, which have been genetically determined, may alter the complex biochemical picture, which one may analyze with the crude technique of studying whole cells. In multicellular organisms the visual measurement of the degree of homozygosity of a strain may furnish useful information. The possibility offered by this technique of revealing genetic differences not evident in the phenotype, as in the case of heterozygotes for recessive mutants, opens new perspectives in the study of dominance relationships, of pleiotropic action of genes, and on the hypomorphic or hypermorphic nature of mutants. The comparison between biochemical differences in the course of development of two or more genotypes producing different morphological traits may give a clue for an understanding of differentiation in chemical terms. For applied genetic work the identification of heterozygotes offers promise, especially in some problems of animal breeding where the diagnosis of carriers of undesired traits gives appreciable economic reward.

At the taxonomic level it is no longer interesting to show that one species differs from another in some biochemical trait. This is what one would expect any way, and it has been abundantly shown that this technique makes this differentiation possible. But there certainly are at least two areas of investigation that appear to offer more fun. By comparing the chromatographic patterns of species belonging to different genera within a family or belonging to different families within a higher taxonomic category one may hope to discover degrees of taxonomic and phylogenetic relationships that are obscure to the morphologist or the cytologist. There are often cases where the systematist has no doubts in establishing relationships between species within a genus, but has no way of deciding whether a genus of a family is likely to be more closely related in phylogeny to this or that other genus or family: in such cases the discovery of common chemical traits may solve the question. Serological methods have also been used for studying the affinities between closely or less closely related forms, our paper chromatography is an alternative method which offers some advantages. Simplicity is the first, but more important is the fact that the zoologist or the botanist may conduct, with the aid of the biochemist, a more precise study of biochemical relationships than with the serological methods, owing to the smaller size of the molecules involved.

Clearly, the biochemical constitution of a cell and of an organism must be the result of the information provided by its desoxyribonucleic acid. One should expect therefore that biochemical analysis would reveal the kind of differences which have been reviewed. Paper chromatographic methods have rendered possible the identification of such differences by means of very simple manipulations, and thus has won their right of citizenship in the domains of genetics and taxonomy. J. B. S. Haldane once wrote [67]: "If I am right in believing that there is at least a correlation between morphological and biochemical diversity, the study of biochemical evolution, based as it must be almost wholly on living organisms, may not be as hopeless a task as has been sometimes assumed." Paper chromatography promises to make this prediction come true.

References

1. BALL, G. H. and CLARK, E. W. *Syst. Zool.* **2**, 138 (1953).
2. BERRIDGE, N. J., CHEESEMAN, G. C., MATTICK, A. T. R. and BOTTAZZI, V. *J. Appl. Bacteriol.* **20**, 205 (1957).

3. BERRY, H. K., CAIN, L. and ROGERS, L. L. *Univ. Tex. Publ.* **5109**, 150 (1951).
4. BERRY, H. K., CAIN, L. and ROGERS, L. L. *Univ. Tex. Publ.* **5109**, 157 (1951).
5. BERRY, H. K., DOBZHANSKY, TH., GARTLER, S. M., LEVENE, H. and OSBORNE, R. H. *Amer. J. Hum. Genet.* **7**, 93 (1955).
6. BLOCK, J. R., DURRUM, E. L. and ZWEIG, G. "A manual of paper chromatography and paper electrophoresis", p. 110. Academic Press, New York (1958).
7. BLOCK, J. R., LE STRANGE, R. and ZWEIG, G. "Paper chromatography. A laboratory manual." p. 195, Academic Press, New York (1952).
8. BOTTAZZI, V. *Ann. Microbiol.* **7**, 3 (1957).
9. BRIMLEY, R. C. and BARRETT, F. C. "Practical chromatography." p. 128, Chapman and Hall, London (1953).
10. BUTZEL, H. M. and MARTIN, W. B. *Genetics* **40**, 565 (1955).
11. BUZZATI-TRAVERSO, A. A. *Nature Lond.* **171**, 575 (1953).
12. BUZZATI-TRAVERSO, A. A. *Proc. Nat. Acad. Sci. Wash.* **39**, 376 (1953).
13. BUZZATI-TRAVERSO, A. A. and RECHNITZER, A. B. *Science* **117**, 58 (1953).
14. CASSIDY, H. G. "Fundamentals of Chromatography. Technique of Organic Chemistry," Vol. 10, p. 235, Interscience Publishers Inc., New York (1957).
15. CHEESEMAN, G. C., BERRIDGE, N. J., MATTICK, A. T. R., BOTTAZZI, V. and SHARPE, M. E. *J. Appl. Bacteriol.* **20**, 195 (1957).
16. CHEN, P. S. *Rev. Suisse Zool.* **63**, 216 (1956).
17. CHEN, P. S. and BALTZER, F. *Nature Lond.* **181**, 98 (1958).
18. CHEN, P. S. and HADORN, E. *Rev. Suisse Zool.* **62**, 338 (1955).
19. CONSDEN, R., GORDON, A. A. and MARTIN, A. J. P. *Biochem. J.* **38**, 224 (1944).
20. CRAMER, F. "Papierchromatographie." p. 135, Verlag Chemie, Weinheim Bergstr (1951).
21. CRUMPLER, H. R., DENT, C. E., HARRIS, H. and WESTALL, R. G. *Nature Lond.* **167**, 307 (1951).
22. CUMMINS, C. S. and HARRIS, H. *J. Gen. Microbiol.* **14**, 583 (1956).
23. DANNEEL, R. and ESCHRICH, B. *Z. Naturf.* **116**, 105 (1956).
24. DANNEEL, R. and ZIMMERMANN, B. *Z. Naturf.* **9**, 12 (1954).
25. DANNEVIG, E. H. *Tidsskr. Hermetikind. No.* **3** (1955).
26. DANNEVIG, E. H. *Fiskeridir. Skr. Havundersøk.* Report on Norwegian fishery and Marine investigations. **11**, 1 (1956).
27. DATTA, S. P. and HARRIS, H. *Eugenics* **18**, 107 (1953).
28. DE GROUCHY, J. and SUTTON, H. E. *Amer. J. Hum. Genet.* **9**, 76 (1957).
29. DE LERMA, B. *Boll. Soc. Ital. Biol. Sper.* **26**, 1 (1950).
30. DE LERMA, B. and DE VINCENTIIS, M. *Boll. Soc. Ital. Biol. Sper.* **31**, 1 (1955).
31. DE LERMA, B. and DE VINCENTIIS, M. *Boll. Zool.* **22**, 1 (1955).
32. DE LERMA, B., DUPONT-RAABE, M. and KNOWLES, M. F. *C. R. Acad. Sci. Paris*, **241**, 995 (1955).
33. DENT, C. E. *Biochem. J.* **41**, 240 (1947).
34. DENT, C. E. *Biochem. J.* **43**, 168 (1948).
35. DENT, C. E. and HARRIS, H. *Ann. Eugen, Lond.* **16**, 60 (1951).
36. DENT, C. E. and ROSE, G. A. *Quart. J. Med.* N.S. **20**, 205 (1951).
37. EGELHAAF, A. *Z. Indukt. Abstamm. -u. Vererb. Lehre* **87**, 769 (1956).
38. FINLEY, H. E. and WILLIAMS, H. B. *J. Protozool.* **2**, 13 (1955).
39. FORREST, H. S. and MITCHELL, H. K. *CIBA Symp.* **143** (1954).
40. FORREST, H. S. and MITCHELL, H. K. *J. Amer. Chem. Soc.* **76**, 5656 (1954).
41. FORREST, H. S. and MITCHELL, H. K. *J. Amer. Chem. Soc.* **76**, 5658 (1954).
42. FORREST, H. S. and MITCHELL, H. K. *J. Amer. Chem. Soc.* **77**, 4865 (1955).
43. FOX, A. S. *Rec. Gen. Soc. Am.* **23**, 40 (1954) and *Genetics* **39**, 967 (1954).
44. FOX, A. S. *Physiol. Zoöl.* **29**, 288 (1956).
45. FOX, A. S. *Science* **123**, 143 (1956).

46. Fox, A. S. *Z. Indukt. Abstamm.-u. VererbLehre* **87**, 554 (1956).
47. Gartler, S. M., Dobzhansky, Th. and Berry, H. K. *Amer. J. Hum. Genet.* **7**, 108 (1955).
48. Goldschmidt, E. *Nature Lond.* **174**, 883 (1954).
49. Graf, G. E. *Experientia* **13**, 404 (1957).
50. Gregory, M. and Mabbit, L. A. *J. Appl. Bacteriol.* **20**, 226 (1957).
51. Gregory, M. and Mabbit, L. A. *J. Appl. Bacteriol.* **20**, 218 (1957).
52. Griffiths, L. A. *Nature Lond.* **180**, 286 (1957).
53. Hadorn, E. *Arch. Klaus-Stift. VererbForsch.* **26**, 470 (1951).
54. Hadorn, E. *Caryologia* suppl. **6**, 326 (1954).
55. Hadorn, E. *Experientia* **10**, 483 (1954).
56. Hadorn, E. "Letalfaktoren in ihrer Bedeutung für Erbpathologie und Genphysiologie der Entwicklung." p. 338. G. Thieme, Stuttgart (1955).
57. Hadorn, E. *Cold. Spr. Harb. Symp. Quant. Biol.* **21**, 363 (1956).
58. Hadorn, E. *Rev. Suisse Zool.* **64**, 338 (1957).
59. Hadorn, E. and Egelhaaf, A. *Z. Naturf.* **11**, 21 (1956).
60. Hadorn, E. and Kühn, A. *Z. Naturf.* **8**, 582 (1953).
61. Hadorn, E. and Kursteiner, R. *Arch. Klaus-Stift VererbForsch.* **30**, 494 (1956).
62. Hadorn, E. and Mitchell, H. K. *Proc. Nat. Acad. Sci.* **37**, 650 (1951).
63. Hadorn, E. and Schwinck, I. *Nature, Lond.* **177**, 940 (1956).
64. Hadorn, E. and Schwinck, I. *Z. Vererbung.* **87**, 528 (1956).
65. Hadorn, E. and Stumm-Zöllinger, E. *Rev. Suisse Zool.* **60**, 506 (1953).
66. Hadorn, E. and Ziegler-Günder, I. *Z. Vererbung.* **89**, 221 (1958).
67. Haldane, J. B. S. "The Biochemistry of Genetics." p. 144. G. Allen and Unwin, London (1954).
68. Hasborne, J. B. and Scherratt, H. S. A. *Nature, Lond.* **181**, 25 (1958).
69. Harris, H. *Ann. Eugen. Lond.* **18**, 43 (1953).
70. Harris, H. "An Introduction to Biochemical Genetics." Eugenics Laboratory mem. 37, p. 96, Cambridge University Press, London (1953).
71. Hoenigsberg, H. F. and Castiglioni, M. C. *Nature, Lond.* **181**, 1404 (1958).
72. Hoenigsberg, H. F. and Castiglioni, M. C. *R.C. Ist. Lombardo. B* **92**, 241 (1958).
73. Hoenigsberg, H. F. and Castiglioni, M. C. *R.C. Ist. Lombardo B* **92**, 249 (1958).
74. Jones, N. R. *Biochem. J.* **58**, 47 (1954).
75. Kaplan, W. D., Holden, J. T. and Hochman, B. *Science* **127**, 473 (1958).
76. Kikkawa, H. *Jap. J. Genet.* **30**, 46 (1955).
77. Kikkawa, H., Ogito, Z. and Fujito, S. *Proc. Jap. Acad.* **30**, 30 (1954).
78. Kirk, R. L., Main, A. R. and Beyer, F. G. *Biochem. J.* **57**, 440 (1954).
79. Koref, S. and Brncic, D. *Drosophila Inform. Serv.* **28**, 126 (1954).
80. Kühn, A. *Naturwissenschaften* **43**, 25 (1956).
81. Kühn, A. and Berg, B. *Z. Vererbung.* **87**, 25 (1955).
82. Kühn, A. and Egelhaaf, A. *Naturwissenschaften* **42**, 634 (1955).
83. Laven, H. and Chen Shen, P. *Z. Naturf.* **11**, 5 (1956).
84. Lee, J. W. *Trans. Amer. Micr. Soc.* **75**, 228 (1956).
85. Lee, J. W., Rene, A. A. and Roddy, L. R. *Trans. Amer. Micr. Soc.* **74**, 311 (1955).
86. Lee, K. Y., Wahl, R., and Barbu, E. *Ann. Inst. Pasteur* **91**, 212 (1956).
87. Merker, E. *Naturwissenschaften* **45**, 118 (1958).
88. Micks, D. W. *Ann. Ent. Soc. Amer.* **49**, 576 (1956).
89. Micks, D. W. and Ellis, J. P. *Proc. Soc. exp. Biol., N.Y.* **78**, 69 (1951).
90. Micks, D. W. and Gibson, F. J. *Ann. Ent. Soc. Amer.* **50**, 500 (1957).
91. McMahon, P. P., von Brant, T. and Nolan, M. D. *J. Comp. Physiol.* **50**, 219 (1957).
92. Nawa, S. and Taira, T. *Proc. Jap. Acad.* **30**, 632 (1954).
93. Neu, R. and Neuhoff, E. *Naturwissenschaften* **44**, 10 (1957).

94. POLONOWSKI, M., JÉROME, M. and GONNARD, P. *CIBA Symp.* **124** (1954).
95. RASMUSSEN, I. E. *Boll. Zool.* **21**, 339 (1954).
96. RASMUSSEN, I. E. *Ric. sci. Conv. gen. Suppl.* **25**, 59 (1955).
97. RASMUSSEN, I. E. and SCOSSIROLI, R. E. *Boll. Zool.* **21**, 329 (1954).
98. RECHNITZER, A. B. *Proc. XIV Int. Congr. Zool. Copen. 1953*, 333 (1955).
99. ROBERTSON, F. W. and FORREST, H. S. *Univ. Tex. Publ.* **5721**, 229 (1957).
100. ROBERTSON, J. G. *Canad. J. Zool.* **35**, 411 (1957).
101. ROGERS, S. *Amer. J. Clin. Path.* **25**, 1059 (1955).
102. ROGERS, S. and BERTON, W. M. *Lab. Invest.* **6**, 310 (1957).
103. SCHÜTTE, H. R., LANGENBEK, W. and BOHME, H. *Naturwissenschaften* **44**, 63 (1957).
104. SCOSSIROLI, R. E. and RASMUSSEN, I. E. *Drosophila Inform. Serv.* **28**, 155 (1954).
105. SCOSSIROLI, R. E. and RASMUSSEN, I. E. *Z. Indukt. Abstamm -u. Vererb.Lehre* **88**, 427 (1957).
106. SHEPARD, C. C. *J. Immunol.* **68**, 179 (1952).
107. SUTTON, H. E. *Eugenics Quart.* **2**, 205 (1955).
108. SUTTON, H. E. and CLARK, P. J. *Amer. J. Phys. Anthrop.* **13**, 53 (1955).
109. STUMM-ZÖLLINGHER, E. *Z. Vererbung.* **86**, 126 (1954).
110. TURBA, F. "Chromatographische Methoden in der Protein-Chemie." p. 358, Spring-Verlag, Berlin (1954).
111. VISCONTINI, M., HADORN, E. and KARRER, P. *Helv. Chim. Acta* **40**, 579 (1957).
112. VISCONTINI, M., KÜHN, A. and EGELHAAF, A. *Z. Naturf.* **11**, 501 (1956).
113. VISCONTINI, M., LOESER, E. and EGELHAAF, A. *Naturwissenschaften* **43**, 379 (1956).
114. VISCONTINI, M., LOESER, E. and KARRER, P. *Helv. Chim. Acta* **41**, 440 (1958).
115. VISCONTINI, M., LOESER, E., KARRER, P. and HADORN, E. *Helv. Chim. Acta* **38**, 2034 (1955).
116. VISCONTINI, M., LOESER, E., KARRER, P. and HADORN, E. *Helv. Chim. Acta* **38**, 1222 (1955).
117. VISCONTINI, M., SCHOELLER, M., LOESER, E., KARRER, P. and HADORN, E. *Helv. Chim. Acta* **38**, 397 (1955).
118. WILLIAMS, R. J. "Biochemical Institute Studies. IV: Individual metabolic patterns and human disease: an exploratory study utilizing predominantly paper chromatographic methods." Univ. Texas Publ. 5109, Austin (1951).
119. WILLIAMS, R. J. "Biochemical Individuality." p. 254, J. Wiley and Sons, London (1956).
120. WRIGHT, C. A., HARRIS, R. H. and CLAUGHER, D. *Nature, Lond.* **180**, 1489 (1957).
121. ZIEGLER-GÜNDER, I. *Z. Vergl. Physiol.* **39**, 163 (1956).
122. ZIEGLER-GÜNDER, I. and HADORN, E. *Z. Vererbung.* **89**, 235 (1958).

TABLE 1

Taxa	Authors and references
ARTROPODA	
BLATTOIDEA	
Blattidea	
Periplaneta americana	Micks, 88; Micks and Gibson, 90.
Periplaneta brunnea	Micks, 88; Micks and Gibson, 90.
Blattelidae	
Blattella germanica	Micks, 88; Micks and Gibson, 90.
Supella supellectilium	Micks, 88; Micks and Gibson, 90.

TABLE 1 (continued)

Taxa	Authors and references
LEPIDOPTERA	
Galleridae	
Galleria mellonella	Robertson, 100.
Tortricidae	
Archips cerasivora	Robertson, 100.
Pyralididae	
Plodia interpunctella	Robertson, 100.
Anagasta kühniella	Robertson, 100.
Lasiocampidae	
Malacosoma distria	Robertson, 100.
Malacosoma fluviale	Robertson, 100.
Ephestia kühniella	Robertson, 100.
COLEOPTERA	
Cucujidae	
Laemophloeus pusilloides	Robertson, 100.
Laemophloeus turcicus	Robertson, 100.
Laemophloeus pusillus	Robertson, 100.
Bruchidae	
Acanthoscelides obtectus	Robertson, 100.
HYMENOPTERA	
Braconidae	
Macrocentrus ancylivorus	Robertson, 100.
Tenthredinidae	
Pristiphora erichsonii	Robertson, 100.
Vespa erabro	Polonowski, Jérome and Gonnard, 94.
LEPIDOPTERA	
Bombyx mori	Polonowski, Jérome and Gonnard, 94.
Pieris rapae	Polonowski, Jérome and Gonnard, 94.
Papilio machao	Polonowski, Jérome and Gonnard, 94.
Opistographis luteolata	Polonowski, Jérome and Gonnard, 94.
Colias croceus	Polonowski, Jérome and Gonnard, 94.
Gonopteryx rhamni	Polonowski, Jérome and Gonnard, 94.
DIPTERA	
Culicidae	
Aedes aegypti	Robertson, 100; Micks and Ellis, 89, Micks 88; Micks and Gibson, 90.
Aedes sollicitans	Micks and Gibson, 90; Micks and Ellis 89.
Aedes quadrimaculatus	Micks and Ellis, 89.
Aedes taeniorhynchus	Micks and Gibson, 90.
Aedes polynesiensis	Micks and Gibson, 90.

TABLE 1 (continued)

Taxa	Authors and references
Aedes varipalpus	Ball and Clark, 1.
Anopheles quadrimaculatus	Micks, 88; Micks and Gibson, 90.
Anopheles albimanus	Micks and Gibson, 90.
Anopheles aztecus	Micks and Gibson, 90.
Culex molestus	Micks and Gibson, 90; Micks, 88.
Culex fatigans	Micks, 88; Micks and Gibson, 90.
Culex pipiens	Micks, 88; Micks and Gibson, 90; Micks and Ellis, 89.
Culex tarsalis	Micks, 88; Micks and Gibson, 90; Ball and Clark, 1.
Culex quinquefasciatus	Micks and Ellis, 89.
Culex salivarius	Micks and Ellis, 89.
Culex stigmatosoma	Ball and Clark, 1.
Culiseta incidens	Micks and Ellis, 89; Ball and Clark, 1.
Culiseta inornata	Micks and Gibson, 90.

Phoridae
Aphiochaeta xantina	Rasmussen, 95.

Calliphoridae
Sarchophaga kellyi	Robertson, 99.

Tachinidae
Drino bohemica	Robertson, 99.

Drosophilidae
Zaprionus vittiger	Rasmussen, 96.
Drosophila robusta	Rasmussen, 96.
Drosophila immigrans	Rasmussen, 96.
Drosophila buzzatii	Rasmussen, 96.
Drosophila repleta	Rasmussen, 96.
Drosophila funebris	Rasmussen, 96.
Drosophila affinis	Rasmussen, 96.
Drosophila algonquin	Rasmussen, 96.
Drosophila obscura	Rasmussen, 96.
Drosophila persimilis	Rasmussen, 96; Scossiroli and Rasmussen, 105.
Drosophila pseudoobscura	Rasmussen, 96; Scossiroli and Rasmussen, 105.
Drosophila subobscura	Rasmussen, 96;
Drosophila melanogaster	Rasmussen, 96; Fox, 44.
Drosophila montium	Rasmussen, 96;
Drosophila simulans	Rasmussen, 96;
Drosphila acanthoptera	Rasmussen, 96; Scossiroli and Rasmussen, 105.
Drosophila busckii	Rasmussen, 96.
Drosophila lebanonensis	Rasmussen, 96; Scossiroli and Rasmussen, 105.
Drosophila victoria	Rasmussen, 96.

TABLE 1 (continued)

Taxa	Authors and references
Drosophila nitens	Rasmussen, 96; Scossiroli and Rasmussen, 105.
Drosophila spinofemora	Rasmussen, 96.
Drosophila setifemur	Scossiroli and Rasmussen, 105.
Drosophila helvetica	Rasmussen, 95.
Drosophila narragansett	Rasmussen, 95.
Drosophila willistoni	Rasmussen, 95.
Drosophila mirim	Rasmussen, 95.
Drosophila baeomyia	Rasmussen, 95.
Drosophila bifasciata	Rasmussen, 95.
Drosophila simulans	Rasmussen, 95.
Drosophila miranda	Scossiroli and Rasmussen, 105.
HEMIPTERA	
Reduvidae	
Triatoma gestaeckeri	Micks, 88; Micks and Gibson, 90.
Triatoma infestans	Micks, 88; Micks and Gibson, 90.
ACARA	
Ixodidae	
Dermacentor andersoni	Micks and Gibson, 90.
Dermacentor variabilis	Micks and Gibson, 90.
Dermacentor albipictus	Micks and Gibson, 90.
ARACNIDA	
Filistata hibernalis	Lee, Rene and Roddy, 85.
Pholcus phalangoides	Lee, Rene and Roddy, 85.
Plexippus paykulli	Lee, Rene and Roddy, 85.
Lacrodectus mactans	Lee, Rene and Roddy, 85.
ECHINODERMATA	
Echinoidea	
Arbacia lixula	Chen and Baltzer, 17.
Sphaerechinus granularis	Chen and Baltzer, 17.
Strongylocentrotus purpuratus	Chen and Baltzer, 17.
Paracentrotus lividus	Chen and Baltzer, 17.
MOLLUSCOIDEA	
Aplexa nitens	McMahon, Von Brant and Nolan, 91.
Australorbis glabratus	McMahon, Von Brant and Nolan, 91.
Austrosuccinia contenta	Kirk, Main and Beyer, 78.
Bothriembryon indutus	Kirk, Main and Beyer, 78.
Bothriembryon kingii	Kirk, Main and Beyer, 78.
Bothriembryon dux	Kirk, Main and Beyer, 78.
Bothriembryon leeuwinensis	Kirk, Main and Beyer, 78.
Helix aspersa	Kirk, Main and Beyer, 78. McMahon, Von Brant and Nolan, 91.
Lanistus baltenianus	McMahon, Von Brant and Nolan, 91.
Lymnaea stagnalis	McMahon, Von Brant and Nolan, 91.

Table 1 (continued)

Taxa	Authors and references
Olata lactea	McMahon, Von Brant and Nolan, 91.
Pomacea zeteki	McMahon, Von Brant and Nolan, 91.
Theba pisana	Kirk, Main and Beyer, 78.

FISHES
Gadidae

Gadus pollachius	Dannevig, 25.
Gadus eallarias	Dannevig, 25.
Gadus minutus	Dannevig, 25.
8 untold species	Dannevig, 25.

Pleuronectidae

Pleuronectes platessa	Dannevig, 25 and 26.
Pleuronectes microcephalus	Dannevig, 26.
4 untold species	Dannevig, 25.

Clupeidae

3 untold species	Dannevig, 25.
Hysterocarpus traski	Buzzati-Traverso and Rechnitzer, 13.
Paralobrax clatratus	Buzzati-Traverso and Rechnitzer, 13.
Paralobrax maculofasciatus	Buzzati-Traverso and Rechnitzer, 13.
Sardinops coerulea	Rechnitzer, 97.

MAMMALIA

Felis catus	Datta and Harris, 27.
Felis leo	Datta and Harris, 27.
Felis tigris	Datta and Harris, 27.
Felis concolor	Datta and Harris, 27.
Felis pardus	Datta and Harris, 27.
Felis bengalensis	Datta and Harris, 27.
Felis serval	Datta and Harris, 27.
Felis pardinus	Datta and Harris, 27.
Felis vivarrina	Datta and Harris, 27.
Genetta tigrina erlangeri	Datta and Harris, 27.
Artictis binturong	Datta and Harris, 27.
Putorius putorius	Datta and Harris, 27.
Canis familiaris	Datta and Harris, 27.
Rattus norvegicus	Datta and Harris, 27.
Mus musculus	Datta and Harris, 27.
Orytolagus cuniculus	Datta and Harris, 27.
Cavia porcellus	Datta and Harris, 27.
Mesocricetus auratus	Datta and Harris, 27.
Macacus rhesus	Datta and Harris, 27.
Equus caballus	Datta and Harris, 27.
Bos taurus	Datta and Harris, 27.
Capra hircus	Datta and Harris, 27.

THE TRANSFER MECHANISMS IN ACTIVE TRANSPORT

W. D. STEIN*

Department of Zoology, King's College, London

INTRODUCTION

ACTIVE transport is the movement of molecules or ions in a direction opposite to that of a prevailing electro-chemical gradient. Without active transport, that is, if left to themselves, molecules or ions move down an electro-chemical gradient, and this tendencey to a uniform distribution of substances tends to obliterate the distinction between an organism and its environment.

An active transport system, however, is one method whereby a cell or organism can maintain a composition different from that of its environment and hence become independent of that environment. A well-explored field in comparative physiology deals with the evolution of the *use* of active transport systems towards the attainment of increasing control of the internal environment [26]. But less attention has been given to the evolution of the active transport systems themselves, and it is to this problem that recent studies on the *mechanism* of active transport may be relevant. Have there been in evolutionary history a number of independent developments of active transport systems or only one, modified for each type of molecule transported? Before we can answer reasonably such a question we will have to know if at the present time there exist a number of fundamentally different types of active transport systems or only one. However, we have at the present, only the beginning of an answer to this preliminary question, as we shall see in what follows.

IDENTIFICATION OF ACTIVE TRANSPORT SYSTEMS

An immediate problem is to determine which molecules are actively transported. To decide whether a permeating molecule or ion is being actively transported we have, clearly, to establish the direction of the prevailing electro-chemical gradient. This is in general a difficult mat-

---

* Present address: Department of Colloid Science, Cambridge, England.

ter, the problem being that the simple *concentration* gradient may bear no relation to the true *electro-chemical* gradient, since a major part of the ion or molecule concerned may not be free to move across the membrane. It may perhaps be adsorbed in the cell interior or constrained in some other way. Thus most of the calcium and magnesium in the interior of muscle cells is generally considered to be bound [5]. Also, the movement of charged particles will be determined by the sum of the prevailing electrical and chemical gradients which may be in opposing directions. Such electrical potentials are of general occurrence in biological systems [14]. In a number of cases, however, active transport in the strict and only correct sense has been convincingly demonstrated. The clearest demonstration is, perhaps, that of the transport of sodium chloride across the isolated frog skin, studied by Huf more than twenty years ago [15] and more recently by Ussing [40]. If the frog skin is placed as a barrier separating two identical "balanced salt" solutions, sodium chloride moves from one compartment into the other. Now, the frog skin cannot be expected to bind significant quantities of the sodium chloride *in the main bulk* of the two solutions. Thus, the measured concentration gradient must be the required electro-chemical gradient and at least one of the two ions concerned must have been actively transported. Whether this was the sodium ion or the chloride ion was not known until Ussing showed that it was possible to neutralize the electrical potential that the frog skin develops, by applying an opposing voltage. Sodium was still transported from one side to the other, in spite of the absence of any electrical gradient. Thus here sodium ions clearly are actively transported—the chloride following passively. In general, such a clear demonstration cannot be expected, for in most cases we cannot control the composition of the solution and the potential on the inner side of cell membranes. Unfortunately, the isolated frog skin preparation, in other ways so useful, may turn out to be unsuitable for a study of the molecular mechanism of active transport, being a complex tissue rather than a homogenous cell membrane.

The Energy Supply for Active Transport

That the organism has to supply energy for the maintenance of active transport is clear. Were it not so, it would be possible to harness the spontaneous diffusion down the electro-chemical gradient to do work, at the expense of no work supplied. In fact, the energy required is derived from cellular metabolism. In a number of cases the metabolic systems responsible for supplying the energy to particular active transport mechanisms is known. Thus it has been established that active transport of sodium and potassium in mammalian red cells is dependent on the anaerobic breakdown of glucose [16] in tortoise red cells on the oxidative pathway of glucose metabolism [18], while in duck red cells and ascites tumour cells, either glycolysis or glucose oxidation can be the energy source [19, 38]. In such cases it is possible, with certain assumptions, to calculate what fraction of the total metabolic energy of the cell goes to maintaining the active transport. This turns out to be a significant portion of the resting metabolism of the cell: in red cells 30% [2], and in ascites cells, 15% [19]. Such figures point to the biological importance to the cell of the maintenance of the active transport system.

Again in the case of red cells, a probably more immediate energy source has been identified, namely adenosine triphosphate (ATP) [37, 41]. ATP being normally unable to penetrate into the red cells, will not, however, support the active transport of sodium or potassium if whole cells are suspended in it. But it is possible to force the ATP into the cells while their permeability barrier is temporarily broken (e.g. during haemolysis) and such internal ATP *will* support active transport [43]. Presumably the ATP-requiring mechanism is situated on the inner surface of the cell membrane.

Morphological Aspects of the Transfer Mechanism

Such knowledge of the energy-supporting reaction, though suggestive, is a long way from a satisfactory explanation of the mechanism of active transport. It may be helpful at this stage to establish what we hope to obtain as such a satisfactory explanation.

What we are concerned with is the *molecular* mechanism of active transport. By this we mean
  i) the identification of those molecules in the membrane directly responsible for the active transport, and
  ii) a detailed description of any chemical or configurational transformations that the transported molecules and the transporting molecules may undergo.

At present, information of this order is available for no transport system but preliminary schemes are beginning to be put forward. (That the attainment of a satisfactory explanation may be a long task is suggested by a consideration of the status of the corresponding problem in muscle contraction. Here the energy-producing system has long been identified and a remarkable body of information collected on the properties of probable components of the contractile system. Yet no generally accepted theory of the part played by these elements in contraction is available. The detailed chemical and spatial transformations of the reacting molecular elements is still controversial, if not unknown [8]).

Let us consider the morphological framework of the problem.

The barrier to the movement of molecules across cell membranes is generally considered to reside in the 50 A° thick bimolecular lipid layer (or half this thickness in bacterial and subcellular components) [7]. It is becoming increasingly evident, however, that the cell membrane is not simply a passive barrier structure but is itself an actively metabolizing organelle. The most complex model that has been put forward is for the yeast cell, where Rothstein has (on physiological evidence) identified a series of three concentric permeability barriers comprising the cell surface, and these barriers and spaces between the barriers are themselves complex and capable of performing a variety of enzymic functions [29]. Similar evidence for the localization of enzymes in cell walls and membranes is available for a number of cell types ranging from bacteria [22] to the intestinal cells of the rat [31].

Such active regions, however, may not pierce the (hypothetical) lipid layer but may merely exist as fairly rigid structures on either side of it. There is not direct observational evidence on this point although such evidence is urgently needed. Thus, in order to account for active transport, we can postulate with some assurance the existence of enzyme and enzyme-like macromolecules at the inner and outer surface of the cell membrane. However, we are on less sure ground when we choose either of the two viewpoints.

(a) that the transported molecule moves through a homogenous lipid layer, or
(b) moves through specially differentiated non-lipidal regions of this layer.

Much discussion has centred around this topic [20]. (Some recent work described by Schulman [32], on the evaporation of water through oil layers suggests that this absolute distinction between a continuous or a non-continuous lipid layer may not, however, be valid. The rate of movement through an oil layer with an adsorbed monomolecular film was greatly modified by fairly slight variation in structure of the adsorbed film. Thus the rate of movement through the lipid layer might be influenced by structures in the non-lipid region. However, since the view that the major permeability barrier is due to the lipid alone is well established [7], more evidence will be needed before these views can be considered as generally applicable.)

Models for the Transport Mechanism

We can now consider some of the current models of active transport systems. Most of these models postulate the presence of a "carrier" in the membrane. This carrier combines with the molecule to be transported to form a complex, which then moves through the membrane. How the complex moves through the membrane and how active transport arises are given various explanations by different authors, as we shall see. Other authors do not postulate such a carrier at all [36].

Phosphate accumulation by bacteria

The most convincing of the carrier models has been put forward by Mitchell and Moyle [22] who have mainly studied phosphate movements across bacterial cell membranes (*Micrococcus* and *Staphylococcus*). These bacterial cells, when actively respiring, can accumulate phosphate against a concentration gradient of at least ten-fold. When respiration is eliminated, however, phosphate does not leak out of the cell. This is not due to an impermeability of the membrane, since experiments with isotopically-labelled phosphate showed that (in the absence of respiration) there was indeed an exchange of phosphate across the membrane. But this exchange was tightly linked so that a phosphate ion could only leave the cell if another phosphate ion entered. Of forty other anions tested, only arsenate could replace phosphate in this linked exchange. The dependence of this exchange of phosphate on the concentration of phosphate indicated that only a limited amount of some membrane component was present. An important finding was that the rate of *uptake* of phosphate during respiration was exactly equal to the rate of *exchange* in the absence of respiration, other factors being constant. Mitchell neatly explained this result and at the same time accounted for the active transport by postulating that the active transport system actually makes use of the exchange-mechanism — uptake occurring by simply shutting off the efflux portion of the exchange. The exchange is mediated, it is assumed, by a carrier system involving enzyme-like "translocases". (These are, Mitchell has suggested, in fact phosphate transferring enzymes.) Efflux is prevented, during respiration, by occupying the carrier system with some metabolic alternative to phosphate — or else by blocking the carrier by the formation of internal chemical bonds.

Ion transport in the red cell

The models that have been proposed to explain ion transport in the red cell have some affinity with Mitchell's model. The relevant experimental data are:
(a) There is a low but definite passive permeability of sodium and potassium across the cell membrane [9, 25].
(b) However, sodium can be extruded and potassium accumulated against a large (fifty-fold) concentration gradient [17].
(c) The active movements of sodium outwards and potassium inwards seem to be tightly linked since:
  (i) there is a strictly observed numerical relation between the number of ions actively moved in and those moved out,

various authors suggesting either a 1 : 1 exchange of sodium for potassium [13] or that 2 potassium atoms are transported inward for 3 sodium transported outward [25] — the constancy of this ratio reported by any one worker is impressive.

(ii) the rate of extrusion of sodium outward is, surprisingly, dependent on the concentration of potassium at the outer face being reduced to very low values as the concentration of external potassium is reduced below 2mM [9].

(iii) conversely, the rate of accumulation of potassium is dependent on the concentration of sodium in the interior of the cell [9].

(d) There can again be demonstrated a saturation of some acceptor component in the membrane, since both the sodium movement outwards and the potassium movement inwards reach maxima as the concentration of the ion concerned is increased [9, 25].

(e) The ammonium ion, lithium and, to some extent, caesium, can replace potassium in that they can be actively transported inwards, competing with potassium for the hypothetical carrier and, in the case of lithium and ammonium, can also maintain sodium transport outwards. No ion can substitute for outward sodium transport [24].

Many of these data have been neatly accounted for by a model suggested by Maizels [17], amplified by Harris [14] and by Shaw [33]. The essential feature of the model is that the same carrier system is concerned in the transport of both sodium and potassium, the carrier being reversibly transformed (in an unknown manner) from a potassium acceptor at the outer face to a sodium acceptor at the inner face. Thus only potassium ions can move in with the carrier and only sodium ions can move out. Metabolic activity is required to transform the carrier from the potassium to the sodium form or vice versa. The carrier is not able to traverse the membrane unless bound to potassium or sodium so that in the absence of either ions, transport ceases and the experimental findings are accounted for. The unsatisfactory feature of the model at present, as is clearly recognized by its propounders, is that not only must we postulate a carrier that can differentiate between sodium and potassium to a greater extent than can familiar chemical systems, but also that this inexplicable affinity ratio must be reversed as the membrane is crossed. However, some recent work has brought the identification of the carrier system closer. Certain pharmacological agents, the cardiac glycosides, have been shown to inhibit this active transport system but, unlike other previously known inhibitors, do so by a direct effect on the carrier system rather than by interfering with metabolism. Glynn has investigated those structural features of the glycosides that confer its inhibitory properties [10], and both he and Wilbrandt [42] have shown that the glycoside acts as a *competitive* inhibitor of potassium uptake. From certain structural features possessed in common by the glycoside and a postulated sterol-potassium complex, Wilbrandt suggests that a sterol molecule is, in fact, the potassium carrier. A consideration of the number of glycoside molecules required for inhibition leads Glynn to calculate that the number of carrier molecules in the membrane is of the order of 1000 per red cell. If we can accept this value, the postulated sterol carrier molecule is likely to form only a very small portion of the total sterol present (10^8 molecules per cell) and identification may prove very difficult.

Amino acid uptake by ascites tumour cells

A somewhat similar approach, using structural chemical features of the transport system in an attempt to define its structure has been used by Christensen and his school [4]. This group has been studying the uptake of amino acids by ascites tumour cells, a particularly useful experimental system as these cells can be obtained in reasonably large quantities of a homogenous preparation of isolated cells. Their capacity to accumulate amino acids is remarkable, concentration gradients of the order of a hundred being obtained in favourable cases. A particular (unnatural) amino acid α, γ-glutamic acid was concentrated so strongly so as to displace much of the potassium of the cells, the cells showing intense swelling of the cytoplasm. Christensen and his group have synthesized a series of amino acids to determine what structural features are particularly favourable for uptake. In addition, following up an observation that vitamin B_6 (pyridoxal) stimulates this uptake, molecules structurally related to the vitamin were tested for their stimulating activity. It has been found that compounds can be formed between these amino acids, pyridoxal and a metal. The stability of these compounds can be correlated with the degree to which the amino acid is concerned in the concentrating abilities of the ascites cells. Thus the plausible suggestion can be made that the carrier molecule in the transport system is related to pyridoxal, and that intermediate compounds are formed during the transfer across the membrane similar to the compounds formed in the test-tube. Here again, an ingenious explanation of a body of experimental data is available, but proof is still lacking of the participation of such pyridoxal-like compounds in the active transport. Pyridoxal is a co-enzyme of amino-transferring enzymes and the possibility exist that, in conformity with Mitchell's theory discussed earlier, the amino acid transport system may be related to a transaminase enzyme with pyridoxal acting as co-enzyme or, in Mitchell's phrase, as a trans-locator. Christensen's group are, however, loath to put forward this speculation on the available evidence, especially since the system concentrates the unnatural D-amino acids and certain rare L-amino acids.

Uptake of ions by yeast cells

Another approach to identification of a carrier system is that of Rothstein and his colleagues who have studied intensively the uptake of sugars [28], phosphate [12] and certain cations [30] by yeast cells. These processes are (strictly) not active transport stystems: the sugars and phosphate are converted into metabolic derivatives during their passage across the membrane, while the cations are probably bound within the cell. Thus transport against a concentration gradient cannot be demonstrated. Nevertheless, these studies have an important bearing upon the present problem. In the first place, as we have seen, Rothstein has firmly established the presence in the yeast cell membrane of many of the enzymes of metabolism and associated with this, has established the structural complexity of this membrane. The more recent results have been equally stimulating to our considerations of the mechanism of active transport. Studies on the uptake of phosphate support a conventional model involving participation of a carrier and saturation of these carrier sites. However, data on the uptake of manganese and magnesium ions, apparently linked to phosphate uptake, have revealed new properties of this carrier system. Apparently, an unstable carrier is metabolically synthesized during phosphate uptake and is thereafter available for uptake of the cations. Thus, while the cells are accumulating phosphate, uptake of cations can occur — but if the completion of uptake of phosphate precedes that of the cations by too long an interval, uptake of cations is slowed down. The postulated cation carrier appears to decay with a half-life of 6 hours at room temperature, in the absence of glucose, but of only 2 hours if glucose is present. The cation carrier is apparently metabolically synthesized during phosphate uptake and then broken down by the cell even when not transporting cations. The specificity of the carrier for cations is reminiscent of that of certain metal-dependent enzymes [30]. Magnesium is, indeed, a co-factor for numerous phosphatases and the suspicion arises here, too, that the carrier may bear some relation to a co-enzyme. More significantly, however, with this information available of the time-course of formation and decay of the carrier, its identification becomes a definite possibility, although as Rothstein points out, it may exist in relatively small quantities.

Attempts at direct isolation of the carrier

The above systems have been chosen from a number of similar systems, since they appear to offer some hope of identification in the near future of parts of the transporting component. There have been

other approaches towards this identification, however. Thus, Solomon and his co-workers have discovered a possible cation carrier substance in blood cells [34]. Extraction of red cells with hot chloroform-methanol mixtures gave a lipid-containing fraction which had in some degree a number of properties to be expected from a cation carrier, namely,
  (a) the ability to take the ions into solution in a non-aqueous medium,
  (b) the ability to liberate these ions again into an aqueous phase, and
  (c) the ability to discriminate between sodium and potassium.

However, these workers showed also that a number of pure lipids (not isolated from red cells) had similar properties and in some cases were better able to discriminate between sodium and potassium. That similar lipids are present in the red cell membrane is perhaps to be expected. Thus Solomon's work reduces to the demonstration that certain lipids possess some of the properties of cation carriers and that such lipids are present in the red cell. The identity of these lipid components with the true carrier remains, as Solomon and his colleagues make clear, to be demonstrated.

Work in our department has also been aimed at the isolation of a membrane transport system [35]. However, we have been studying, not true active transport processes, but so-called facilitated diffusion systems. These are concerned in the acceleration of movement down-the-concentration-gradient of certain non-electrolytes, glycerol and glucose [3]. It is generally considered that these facilitated diffusion systems may show certain features in common with active transport mechanisms. Facilitated diffusion has been accounted for in terms of a carrier theory, but other models involving absorbing surfaces and protein-lined pores have also been suggested. We ourselves have previously favoured the latter view. However, recent work suggests that a carrier model may be more valid. The glycerol-facilitated diffusion system is inhibited by traces of copper and we have been able to show that the copper-binding group of the system is an amino-acid (histidine) possessing a free amino group. If not all, then the major part of this so-called N-terminal histidine is concerned in glycerol transport. We had interpreted this as evidence for a protein pore, the histidine being apparently at the end of a protein chain. We now find, however, that the N-terminal histidine under certain circumstances can be brought into an acetone-soluble form, suggesting that it may be linked to a lipid moiety. The amount of histidine present in the membrane suggests, with certain assumptions, that it is part of a fairly small molecule having a molecular weight not much greater than 1000. Such a molecule could be a phospholipid with histidine attached to the serine of a cephalin molecule. (Similar cephalins containing amino-acids other than serine have recently been identified.) This suggestion as to the moiety to which the N-terminal histidine is attached is purely tentative, however, although work is in progress to test this point. The role of histidine in the catalytic activity of enzymes is becoming increasingly evident [1] and a number of suggestive carbohydrate-histidine interactions have been postulated. Histidine in enzymes appears also to act as a reversible acceptor of phosphoryl and acetyl groups. Plausible models of facilitated diffusion can no doubt be suggested on this type of information. It is our hope, however, that the molecular structure of the glycerol carrier will soon be established, so that such speculations can be placed on a firmer footing.

Classification of Active Transport Systems

It may be worthwhile to consider the postulated active transport models more systematically. We can distinguish, in the first place, between those mechanisms in which the membrane component (or part of it) is mobile and those in which all the membrane components are fixed. Let us call these mobile carrier and fixed carrier models. Some border-line models will be more or less "fixed". Although a completely non-mobile carrier model can be a plausible model for some type of facilitated diffusion, it does not seem easy to account for active transport on such a scheme [27] and we need consider it no further here.

Considering then the mobile carriers, we have generally, a model whereby the mobile membrane component is able to interact with the transported molecule at either side of the membrane and to cross the membrane in a manner which at the moment we can ignore. Now, a good way to organize active transport is, as in Mitchell's model, to ensure that the carrier is only available to transport the permeant in one preferred direction—from the low to the high concentration area. This type we may term the "inhibited carrier" model. How the carrier is rendered differentially available will vary from case to case. Thus, for example, the with-the-gradient movement of carrier may occur only in conjunction with an inhibitor, or else only in the absence of an activator. The essential features are this differential availability and the fact that the movement of free and combined carrier is equally allowed in both directions. In a second general type of active transport mechanism this is not the case. These are the "energized carrier" models. (Patlak has presented an example of this type of carrier [23].) The carrier is able to combine freely with the molecule to be transported at both sides of the membrane. Active transport arises from an active increase in velocity of movement of the carrier in the direction of the electro-chemical gradient. Alternatively, a similar increase in the velocity of the carrier-transportant complex, when moving in the opposite direction, will also give rise to active transport. For instance, modifying a model of Danielli's [6] we may have as the carrier a contractile protein energized by an enzyme localized at one side of the membrane only. The molecule or ion to be transported is an activator (in the biochemical sense) of this enzyme. When the required molecule, say the potassium ion, impinges on the system, the enzyme is activated —the protein contracts—and the potassium moves through the membrane as in the figure. The protein can also move through the membrane passively, either free or when combined with the transportant,

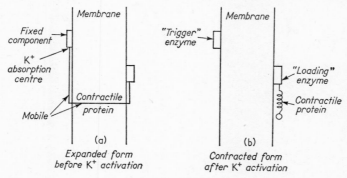

Fig. 1.

but this movement is slow. The increase in velocity occurs only when the transportant impinges on the preferred site and sets off the energizing reaction. Such a system will result in the accumulation of the activating molecule or ion at the expense of metabolite. (An earlier similar model of Goldacre [11] was a fusion of both the above types of models, movement of the carrier being enhanced by a contractile protein but also the carrier binding transporting differentially at the two sides of the membrane.)

Now a fundamental difference between these two general mechanisms is their response to the removal of the energy-producing systems, i.e., to the absence of respiration or metabolism. In the "inhibited carrier" model the total movement of transported molecules reaches a maximum in the absence of respiration. This is because the total exchange reaches a maximum—molecules being ferried in *both* directions. On the other hand, in the "energized carrier" model the total movement reaches a minimum, movement reducing to the low level of the free diffusion of carrier and transportant. Just this difference is found between the two most intensively studied systems: the phosphate transport system in bacteria [21] and the ion transport system in the red cell. In this latter case exchange is reduced to a low level in the absence of respiration [17]. It is possible, therefore, that these two systems involve two fundamentally different types of mechanism. Two more points arise from this distinction between the two types of active transport mechanisms. First, the inhibited carrier type, but not the energized carrier type, can act both as facilitated diffusion and as active transport mechanisms. For it will be noted that the inhibited carrier type does not require an energy input in order to allow exchange to occur. Any ontogenetic or phylogenetic relation between active transport and facilitated diffusion phenomena must therefore be restricted to the inhibited carrier mechanisms. Second—and even more tenuously—in the models we have been considering, the relation between transported molecule and carrier in the inhibited carrier type is a fairly passive one and analogous to that between substrate (or competitive inhibitor) and enzyme. On the other hand, in the energized carrier type, the transported molecule causes a rather violent change in the properties of the membrane component and the analogy is more between activator (or non-competitive inhibitor) and enzyme. We can say, loosely, that an enzyme acts upon its substrate, whereas an activator acts upon its enzyme. In the former case the enzyme (or membrane component) is not itself much altered. In the latter case there is a marked change in the properties of the enzyme. Now the cations are more commonly enzyme activators and inhibitors than substrates, the larger organic chemical molecules more commonly substrates. It may be, therefore, that the energized carrier type is more confined to the active transport of the cations (and perhaps chloride) and in fact has evolved from some enzyme-activator system—whereas the inhibited carrier type is more confined to transport of the non-mineral metabolites and has evolved from the enzyme-substrate systems.

Conclusions

It is clear that a number of suggestions as to the mechanisms of action of active transport systems are available. These have had the virtue, as Ussing remarks [39], of convincing the scientific public that active transport is "thermodynamically possible and not at all mysterious." It is now necessary vigorously to attack the problem of determining just which of the postulated models is in fact operative in any active transport system.

The immediate need is for the unambiguous identification of those membrane components responsible for active transport. Important too, is the necessity for firmly establishing the model system in a realistic physico-chemical picture. We need physico-chemically plausible models for carrier transport. For this we need information on the types of movement that molecules in interfaces can undergo. Recent work by Schulman's school on the permeability of oil layers [32] is likely to offer the type of answer that is required. Finally, much accurate information on the spatial arrangement of components of the membrane is required. Here electron-microscopic studies at the cytochemical level will be of value but also the type of analysis that Rothstein has made on the yeast cell [29] and Mitchell on the bacterial cell [21] need to be extended to other cell types. With this morphological information together with a knowledge of the structure and chemical properties of the active transport components and, finally, with a clearer physico-chemical picture of membrane organization, we can expect a realistic account of the mechanism of active transport.

References

1. BARNARD, E. A., and STEIN, W. D. *Advanc. Enzymol.* **20**, 51 (1958).
2. BERNSTEIN, R. E. *Nature, Lond.* **172**, 911 (1953).
3. BOWYER, F. *Inter. Rev. Cytol.* **6**, 469 (1957).
4. CHRISTENSEN, H. N. and RIGGS, T. R. *J. Biol. Chem.* **220**, 265 (1956).
5. CONWAY, E. J. *Physiol. Rev.* **37**, 84 (1957).
6. DANIELLI, J. F. *Symp. Soc. Exp. Biol.* **6**, 1 (1952).
7. DAVSON, H., and DANIELLI, J. F. "The Permeability of Natural Membranes." Cambridge (1952).
8. EDSALL, J. T. and others Symposium on Biocolloids, *J. Cell. Comp. Physiol.* **49**, Supp. 1 (1957).
9. GLYNN, I. M. *J. Physiol.* **134**, 278 (1956).
10. GLYNN, I. M. *J. Physiol.* **136**, 148 (1957).
11. GOLDACRE, R. J. *Inter. Rev. Cytol.* **1**, 135 (1952).
12. GOODMAN, J. and ROTHSTEIN, A. *J. Gen. Physiol.* **40**, 915 (1957).
13. HARRIS, E. J. *Symp. Soc. Exp. Biol.* **8**, 228 (1954).
14. HARRIS, E. J. "Transport and Accumulation in Biological Systems." Butterworths, London, (1956).
15. HUF, E. G. *Arch. Anat. Physiol. Lpz.* **237**, 143 (1935).
16. MAIZELS, M. *J. Physiol.* **112**, 59 (1951).
17. MAIZELS, M. *Symp. Soc. Exp. Biol.* **8**, 202 (1954).
18. MAIZELS, M. *J. Phsyiol.* **132**, 414 (1956).
19. MAIZELS, M., REMINGTON, M. and TRUSCOE, R. *J. Physiol.* **140**, 80 (1958).
20. MITCHELL, P. *Disc. Faraday Soc.* **21**, 278 (1956)—and other contributors to this discussion.
21. MITCHELL, P. *Nature, Lond.* **180**, 134 (1957).
22. MITCHELL, P. and MOYLE, J. *Biochem. J.* **64**, 19P (1956).
23. PATLAK, C. S. *Bull. Math. Biophys.* **19**, 209 (1957).
24. POST, R. L. *Fed. Proc.* **16**, 442 (1957).
25. POST, R. L. and JOLLY, P. C. *Biochim. Biophys. Acta* **25**, 118 (1957).
26. PROSSER, C. L. "Comparative Animal Physiology." W. B. Saunders & Co., Philadelphia (1950).
27. ROSENBERG, T. *Acta Chem. Scand.* **2**, 14 (1948).
28. ROTHSTEIN, A. *Symp. Soc. Exp. Biol.* **8**, 165 (1954).
29. ROTHSTEIN, A. *Disc. Faraday Soc.* **21**, 229 (1956).
30. ROTHSTEIN, A., "Proc. U.N. Atoms for Peace Conf." (Geneva), (to be published).
31. ROTHSTEIN, A., MEIER, R. C. and SCHARFF, T. G. *Amer. J. Physiol.* **173**, 41 (1953).
32. SCHULMAN, J. H. *Disc. Faraday Soc.* **21**, 284 (1956).

33. Shaw, T. I., quoted in Glynn, I. M. *Progr. Biophys.* **8**, 241 (1957).
34. Solomon, A. K., Lionetti, F. and Curran, P. F. *Nature, Lond.* **178**, 582 (1956).
35. Stein, W. D., *Nature, Lond.* **181**, 1662 (1958).
36. Stein, W. D. and Danielli, J. F. *Disc. Faraday Soc.* **21**, 238 (1956).
37. Straub, F. *Acta Physiol. Hung.* **4**, 235 (1955).
38. Tosteson, D. C. and Robertson, J. S. *J. Cell. Comp. Physiol.* **47**, 147 (1956).
39. Ussing, H. H., in Clarke, H. T. (ed.), "Ion Transport Across Membranes." p. 10. Academic Press, New York (1953).
40. Ussing, H. H. and Zerahn, K. *Acta. Physiol. Scand.* **23**, 110 (1951).
41. Whittam, R. *J. Physiol.* **140**, 479 (1958).
42. Wilbrandt, W. *J. Pharm. Lond.* (to be published).
43. Gárdos, G. *Acta Physiol. Hung.* **6**, 191 (1954).

THE MATCHING OF DRUGS TO TUMOURS

P. Hebborn

Department of Zoology, King's College, London

From both an etiological and a morphological point of view, cancer represents a large group of diseases possessing as common characteristics a capacity for uncontrolled proliferation and a variable amount of cellular dedifferentiation. Tissue culture experiments illustrate that these properties belong to any normal tissue under appropriate culture conditions. However, this process is reversible, since such cells can be made to revert to type by placing them under the normal controlling forces of the body. Until more is known of these controlling mechanisms and of the neoplastic transformation, the use of drugs that will destroy the cancer cell as opposed to restoring it to its former non-cancerous condition, offers the main means of therapy, particularly in those cases where surgery and radiation are not possible.

Any chemical having its greatest cytotoxic effect against dividing cells is a potential anti-cancer agent. The majority of drugs interfere with cell division by affecting the mechanics of cell division, by preventing the synthesis of tissue components, by interfering with energy-producing metabolic processes or by other methods. Drugs having these effects on a cancer cell also have the same effects on every other dividing cell in the body. Excluding hormone therapy, the apparent selectivity of action of those drugs used clinically depends upon their being used against the more rapidly growing neoplasias, e.g. leukaemia, which are susceptible to attack because their rate of mitosis is equal to or in excess of that for the normal rapidly dividing tissues in the body, e.g. intestinal mucosa and skin.

In other fields of chemotherapy, specificity of action is obtained by utilizing marked differences in metabolic behaviour between the host and the infecting organism: tumour cells closely resemble many normal cells which cannot tolerate extensive damage. However, it is theoretically possible to make use of even small differences between normal and cancer cells in designing tumour-specific drugs. Various ways of increasing the specificity of one series of anti-tumour agents, the nitrogen mustards, may be envisaged.

The anti-mitotic effect of nitrogen mustards depends upon the

TABLE 1

1 Drug	2 % Hydrolysis	3 Effect on tumour	4 Activating enzyme in tumour
R = ⟨⟩-N(CH$_2$CH$_2$Cl)$_2$			
NH$_2$R	>100	+	
CH$_3$CO.NH.R	42	++	+
C$_6$H$_5$CO.NHR	40	−	−
CCl$_3$.CO.NHR	13	++	+
CH$_2$F.CO.NHR	19	+	+
CF$_3$.CO.NHR	7	++	+
OH.R	60	++	
CH$_3$COO.R	15	+++	+
C$_6$H$_5$.CO.O.R	15	++	+

chemical reactivity of the molecule, i.e. the ease of ionization to form carbonium ions:

$$R-N\begin{matrix}CH_2CH_2Cl\\CH_2CH_2Cl\end{matrix} \rightleftharpoons R-N\begin{matrix}CH_2CH_2Cl\\CH_2CH_2^+\end{matrix} + Cl^-$$

Substituting electron-attracting groups into the molecule reduces the ionization rate, rendering the molecule inactive to an extent dependent upon the substituent group. The biological activity of a nitrogen mustard substituted in this fashion depends upon two independent variables, (a) the ability of a tissue to remove the inactivating group by enzyme action, and (b) the mitotic index of the tissue.

For example,

$$CCl_3 \cdot CO.NH \langle \rangle M \xrightarrow{acetylase} NH_2 \langle \rangle M + CCl_3COOH.$$
$$\text{inactive} \qquad\qquad\qquad \text{active}$$

where $M = {}^-N(CH_2CH_2Cl)_2$.

The greater number of variables on which a drug depends for its activity, the more selective will that drug be [2].

Preliminary experiments have been conducted to test this theory using various derivatives of I and II.

$$NH_2 \langle \rangle M \quad I$$
$$OH \langle \rangle M \quad II$$

A study of the effect of these mustards and various acylated derivatives on the Wistar rat and the Walker carcinoma growing subcutaneously indicate that, in general, the derivatives have a greater anti-tumour effect and a reduced systemic effect compared with the parent compound. These results are partly summarized in Table 1 and are reported elsewhere [3]. The hydrolysis rates in column 2 are a measure of the chemical reactivities of the molecules. Although the derivatives have a reduced chemical reactivity, the majority of them have an increased anti-tumour effect. This can be correlated with the presence of an activating enzyme in the tumour. The results obtained using such simple acylating groups, which are completely foreign to the normal metabolism of the cell, indicate that a greater degree of specificity may be expected from the use of derivatives designed to utilize the following properties of cancer cells.

Protein and Nucleic Acid Synthesis

Most tumours are able to concentrate the necessary material for protein and nucleic acid synthesis in competition with other tissues in the body. It is possible that the building materials for these substances enter the cancer cell by means of
(a) an altered permeability of the cell wall,
(b) a facilitated diffusion mechanism,
or (c) an active transport mechanism [4].
It is possible that derivatives possessing a purine, pyrimidine or amino acid moiety may be selectively absorbed into cancer cells by virtue of their anabolically important groups fitting one of these mechanisms. For example, III is a modified adenine molecule, IV and V are substituted cytosine and orotic acid molecules and VI is an arginine molecule which could be substituted in a number of positions to carry the mustard molecule.

$$\text{V} \quad \underset{\text{OH}}{\underset{|}{\text{N}}}\underset{\text{N}}{\overset{\text{OH}}{\bigcirc}}\text{CO.NH}\text{–}\langle\text{–}\rangle\text{–N}\begin{smallmatrix}\text{CH}_2\text{CH}_2\text{Cl}\\ \text{CH}_2\text{CH}_2\text{Cl}\end{smallmatrix}$$

$$\text{VI} \quad \text{NH}=\text{C}\begin{smallmatrix}\text{NH}_2\\ \text{NH}\end{smallmatrix}\quad \text{etc.}$$

(Structures V, VI, VII shown)

V: 2,4-dihydroxypyrimidine-6-carboxamide linked to phenyl-N(CH$_2$CH$_2$Cl)$_2$

VI: guanidino–(CH$_2$)$_3$–CH(NH.CO-C$_6$H$_4$-N(CH$_2$CH$_2$Cl)$_2$)–CO.NH-C$_6$H$_4$-N(CH$_2$CH$_2$Cl)$_2$

VII: COOH.$\overset{*}{\text{C}}$H.CH$_2$–⟨–⟩–N(CH$_2$CH$_2$Cl)$_2$ with NH$_2$

It is probable that VII, an amino acid derivative, enters this class of compounds, because the optically active isomers differ remarkably in their anti-tumour effects although their chemical reactivities are the same. The D-isomer is non-inhibitory, the L-isomer has marked anti-tumour activity and the DL-isomer has intermediate activity [5].

Invasive Properties

It seems probable that tumours produce a "spreading factor" which assists the invasiveness of tumours by breaking down connective tissue barriers. Biochemical investigations reveal a low activity of the hyaluronidase complex of enzymes in tumours. Histochemical methods may demonstrate the localization of this or another enzyme complex in the region of invading cancer cells. If so, then a derivative substituted with a glucoside, ester, acetylamino or aldehydamino linkage may be selectively activated by the more invasive cancers.

High Enzyme Activity Values in Tumours

Although the neoplastic transformation is associated with a general decrease in most enzyme activity values, some enzymes, e.g. some intracellular peptidases, are reported to have an increased activity in tumour tissue compared with the tissue of origin. Also, tumours of the melanoma type should be susceptible to attack using drugs which could be activated by the enzymes concerned in melanin deposition.

It is unlikely that a single anti-mitotic agent will be discovered that will selectively attack all types of cancer. However, considerations such as the above may produce drugs that are more selective for various tumours than those that exist at the present time, not only in the nitrogen mustard series, but in other anti-mitotic series of compounds.

References

1. Ross, W. C. J. "Advances in Cancer Research" Vol. **1**, 397, Academic Press Inc., New York (1953).
2. Danielli, J. F. *Nature, Lond.* **170**, 863 (1952).
3. Hebborn, P., and Danielli, J. F. *Biochem. Pharmacol.* **1**, 19 (1958).
4. Danielli, J. F. *Ciba Symp. Leukaemia Research* p. 263 (1954).
5. Bergel, F. and Stock, C. C. *Rep. Brit. Emp. Cancer Campgn.* **31**, 6 (1953).

THE CYTOCHEMISTRY OF PROTEINS

E. A. BARNARD

Department of Zoology, King's College, London

For well over one hundred years, proteins have been isolated from animal tissues and studied in that state with increasing precision. There has long been an awareness, still growing, of their immense significance in all cellular processes and situations. But only recently has a start been made on the task of studying with comparable precision the nature, distribution, properties and activities of proteins in their original state in animal tissue. This not unambitious aim is the programme of protein cytochemistry.

It is easy to demonstrate the real need for specifically cytochemical methods here. It now seems clear, from much evidence, that a large number of biological mechanisms depend at the cellular level upon a specific organization over very small distances. To investigate these, a high degree of resolution in space is required, attainable only by microscopic methods. Further, the cells in any one tissue form a heterogeneous assembly. Differences among them in protein distributions and functions will often correspond to important biological differences. Only *in situ* cytochemical methods can give the discrimination required for this purpose.

There are two basic classes of information on proteins in tissues that we can derive cytochemically:

(A) information on different species of protein macromolecules
(B) information on different protein side-chain groups, e.g. –SH groups, tyrosine phenolic groups, etc.

Dealing with class B first, we might, in principle, find a method to measure all the groups of a given type R_1, in any particular region, e.g. all the –SH groups irrespective of the parent molecules to which they belong. There are several reasons why we might seek such a method. One is that this is normally the only way that we can study such R_1 groups. That is, there is no method known, even in studies on isolated crystalline proteins, of distinguishing between, say, two –SH groups solely by virtue of their being on two different types of protein molecule. Another reason is that the fact that R_1 groups occur on different protein molecules, which can be separated biochemically, will often be less significant than the fact that a particular local assembly of R_1 groups occurs in the cell. As one example among many, a sequence of changes in the amount of –SH groups has been claimed to be of importance in mitosis [1], and one should be able to examine this cytochemically.

The methods available for protein cytochemistry of this type are
(1) Staining with dyes.
(2) The use of intrinsic light absorption spectra.
(3) The use of reagents selective for particular chemical groups.

The century since Darwin has seen the rise of the dye-staining methods in cytology, and now, probably, their fall. They have been, and still are, of great value in purely histological applications. But the interactions of dyes with cell components are *adsorptions* of various types; they are determined by the local surface phases and not in any direct manner by the simple presence of particular protein side-chain groups. The dye bound is in equilibrium with dye in the bulk phase, and is thus removable to a greater or lesser extent by washing. The interactions of dyes with proteins will be modified by the presence of polynucleotide, carbohydrate or lipid elements. For these and similar reasons, the dye-binding methods cannot yield true cytochemical information, other than of a very generalized and approximate kind.

Method 2 is restricted in scope. Suitable absorption is shown only by the side-chains of tyrosine, phenylalanine and tryptophane [2], and by certain protein prosthetic groups, e.g. haem [3].

The essential characteristic of methods of type (3) is that true covalent chemical bonds are formed. The aim is to develop a selective absorption of light at the sites of the group in question in the protein molecule. The ideal requirements for a reagent of this type are as follows:

(i) Stable covalent bonds should be formed at the groups concerned.
(ii) (a) The reaction should be selective for a particular group or groups.
(b) It should be known with which groups in the tissue specimen the reagent will react, and to what extent the reaction is quantitative.
(iii) The reagent should not give rise to adsorption artifacts, e.g. it should be colourless and readily removed by washing with inert solvents.
(iv) No other disturbing modifications of the tissue components should be produced by any of the treatments used.
(v) The reaction product should have an absorption spectrum and an extinction coefficient suitable for accurate microscopic observation.

These requirements are, in practice, quite stringent. The reactivities of different groups in proteins overlap, and it is difficult to find a suitable reagent that will react at only one type of group to form a sufficient colour. For this reason, blocking reactions were proposed by Danielli [4], i.e. the range of groups that react with a particular colour-forming reagent is reduced by prior treatment with a reagent that reacts with only some of them to give a colourless product—this latter is a blocking reagent.

The diazonium coupling reaction may be taken as an illustration of methods of this type. The reaction of diazonium reagents with proteins to form coloured products has long been known, and has been applied in much immunological work. A suitable cytochemical application [5] involves treatment firstly with tetrazotized benzidine and secondly with a naphthol. Thus, at a protein tyrosine group:

$$\text{Protein}-\text{C}_6\text{H}_4-\text{OH} \rightarrow \text{Protein}-\text{C}_6\text{H}_3(\text{N}=\text{N}-\text{C}_6\text{H}_4-\text{C}_6\text{H}_4-\text{N}_2^+ \text{OH}^-)-\text{OH} \xrightarrow{+\beta-\text{naphthol}}$$

$$\text{Protein}-\text{C}_6\text{H}_3(\text{N}=\text{N}-\text{C}_6\text{H}_4-\text{C}_6\text{H}_4-\text{N}=\text{N}-\text{C}_{10}\text{H}_6\text{OH})-\text{OH}$$

Here, a strong final colour is produced due to the bis-azo dye formed, no coloured compound is applied to the tissue, excess reagent can be washed away at each stage, and the naphthol component used can be altered to check on adsorption artifacts. This treatment results in a strong overall reaction throughout a tissue section, which in fact approximately indicates the distribution of total protein. The main sites of reaction are protein tyrosine and histidine residues, while small contributions to the visible colour may come from tryptophane and certain other groups.

However, when benzoyl chloride is used as a prior blocking agent, the coupling colour is observed, in general, only in the cell nuclei [6]. Detailed analysis of this reaction has shown it to be due to certain histidine groups in a special situation in the nucleoprotein of the nuclei. Thus, a range of reactive groups has been narrowed down by prior blockage to leave available for reaction only one particular set, and useful information on this set of histidine groups can now be obtained cytochemically.

In general, experience in this field now shows that for the study of protein groups, attention should be concentrated on (i) the use of suitable blocking reaction sequences, or, better still, (ii) the design of single reagents completely specific for a given group, and (iii) determination by various techniques of micro-analysis what groups in the tissue specimen have in fact reacted with a particular reagent.

I turn now to methods of class A, giving localizations of entire protein molecules. In practice, these molecules must be defined by their activity. The methods available are:

1. Localizations of enzymic activities
2. Reactions using labelled antibodies.

1. *Enzyme methods.* The enzyme in the specimen acts on a substrate, and the aim is so to arrange the reaction conditions that a coloured end-product is formed and remains at the same site. This may be done, firstly, by *trapping* the reaction product, i.e. a reaction occurs to precipitate one of the products as it is liberated. Thus, when a phosphatase acts on a phosphate ester, the phosphate ions liberated can be precipitated by calcium ions present in the solution, and the deposits of calcium phosphate can be rendered visible by exchange with cobalt and reduction to cobalt sulphide. Similarly, the organic molecules liberated in various enzymic hydrolyses can be trapped, e.g. by oxidation to a dye in the methods of Holt [7] using indoxyl esters.

Secondly, methods have been studied where the initial product itself is insoluble—this is of limited applicability.

Thirdly, for oxidative enzymes, *acceptors* are used which, during the reaction of the enzyme with a substrate, are converted by reduction to an insoluble, coloured form, e.g. tetrazolium salts which on reduction give formazans.

It is important to note that with any of these methods, a covalent bond is not formed onto the enzyme. The deposition is due to insolubility and may be aided by adsorption. There is much opportunity for diffusion and false localizations in such a system, and very careful control of the reaction is essential.

Information on enzymes within cells can be one of the most important contributions of cytochemistry to cell biology; but owing to the nature of the techniques that we rely on at present, each individual method must be carefully scrutinized for reliability.

2. *Antibody methods.* Proteins are normally antigenic, and if antibodies are obtained to a specific cellular protein, and applied to the tissue section, they will be specifically and firmly precipitated at the sites occupied by that protein. In this method, introduced by Coons [cf. 8], the antibody molecules are made fluorescent by previous reaction with a suitable compound; the sites of deposition are thus visible in the fluorescence microscope. This can be a powerful method, since use is

made of the immunological discrimination between different proteins, which is very sensitive to small changes in structure. Also, fluorescent markers are much more sensitive than normal light absorption markers. There are difficulties with regard to certain aspects of this method, but in principle it can allow the localization of almost every cell protein.

Limitations

Having outlined the methods available, it must be pointed out that there are certain general difficulties in protein cytochemistry, which must be fairly central to any consideration of what can be achieved with this discipline.

(1) *Limitations inherent in the nature of tissue protein systems*

It is necessary to stress the great complexity of proteins as physicochemical systems. Our knowledge of the detailed structures of proteins, or even of their amino acid sequences alone, is still for the most part meagre. The complexities introduced by their size and their charge distributions, and by their multiple configurations, add to the effects of interactions between different chemical groups in the protein structure.

Chemical studies of proteins are, not surprisingly, still far from enabling us to analyze and evaluate these effects, which must be fundamentally related to the biological activities of the proteins. Hence protein cytochemistry must operate under a disadvantage until we have a more sophisticated knowledge of protein macromolecules. The situation is still more difficult with regard to proteins in tissue specimens where various, largely unknown, combinations with other proteins and with lipid, carbohydrate and nucleotide molecules will add further complications to these effects.

This type of difficulty must be approached now by empirical means. Thus it is dangerous to interpret the results of cytochemical experiments using reactivities obtained from simple organic chemical considerations. Many cases are now known where the reactivity of a particular amino-acid side-chain group in various situations differs very considerably from that of the parent amino-acid (cf. for some examples, reference 9).

Therefore, in methods aiming at particular side-chain groups, the reactivity of a free amino-acid towards the reagent is an insufficient guide. It should be supplemented by examining this reactivity in peptide and protein systems, and evidence should be sought from the results obtained with alternative or complementary reagents to support the cytochemical interpretations. Even so, this will often be inadequate to interpret reactions in tissue specimens; thus, in the example of the diazonium coupling of histidine residues referred to earlier, it is not predictable that, while histidine residues in the cytoplasm would react normally with benzoyl chloride, some nucleoprotein histidine residues would fail to do so, although all are reactive to the diazonium reagent. The best approach at present for establishing a cytochemical interpretation is to determine by micro-analytical methods on reacted specimens which chemical groups have reacted in a given case. Modern spectroscopic and chromatographic methods can render this task easier.

(2) *Limitations introduced in the treatment of the material*

For cytochemical observation it is necessary to render the component studied immobile at its original site, and to bring it into a condition suitable for optical examination. This, of course, normally means that fixation must be used. We still have an incomplete understanding of the chemical and physical changes produced in proteins when a tissue is fixed, but it is known that they are often considerable. This can raise difficult problems for cytochemistry. Thus, we may reasonably wish to determine the total amount of a given protein group *or* the amount of it accessible in the native state. Even if all other conditions are ideal, what we may determine in practice is the amount of this group that the fixative used has made available to the reagent—which may often be a quantity of little physiological interest.

Similar problems arise in the other methods. Thus, different enzymes have been found to be inactivated to varying degrees by fixatives. Similarly, changes in proteins on denaturation may change their antigenic properties. In all cases, an empirical study has to be made of the effects of fixation on the component being examined. Analytical methods can be applied to show the amounts of enzyme activity destroyed in the procedures used, but in those cases where this destruction is high, the difficult problem remains of finding satisfactory methods for preserving the activity consistent with its cytochemical demonstration.

Another type of difficulty arising in the treatment of the specimen is that due to diffusion. Diffusion in fixation can be very greatly reduced by using the Altmann-Gersh freeze-drying method. The advantages of this method in comparison with chemical fixations, e.g. the very small extent of diffusion of macromolecules, the absence of chemical modification of cell components by reaction with the fixative, and the high quality of preservation of cytological detail, indicate that freeze-drying should be used whenever possible in accurate cytochemical work. Subsequent denaturation is still required here. However, the action of a fixative (such as absolute ethyl alcohol) on the macromolecular matrix present in the frozen-dried specimen, in practice produces much less disturbance in the components than occurs in the mobile systems which are present when a fixative diffuses into fresh tissue. We may be unable to escape the necessity for using fixation in one form or another, but our difficulties here arise from insufficient knowledge of the relation that the fixed specimen bears to the native one.

Measurement. There is a growing emphasis now in cytochemistry on a search for quantitative methods, as these are essential in any serious application. There are large errors possible in an evaluation of a stain by eye, even for very approximate assessments. Problems exist in transferring quantitative spectrophotometry to the microscopic level, but progress has been made in recent years in this respect, especially in connection with nucleic acid cytochemistry [cf. 10] — a field where in some ways the material is more tractable than in the protein case. We should now expect an increasing demand that our methods meet the requirements of quantitative micro-spectrophotometry.

Another approach to the problems of measurement which is now being successfully applied, and which holds promise for many enzymic reactions, is that of interference microscopy. The technical aspects of this method will be dealt with elsewhere [11], so that it is only necessary to add that the dry mass determinations have been applied to measure the amount of a precipitate deposited by enzyme activity e.g. the calcium phosphate formed with alkaline phosphatase [12]. The increase of mass with time is followed at a given site; the initial slope of this curve is a measure of the enzyme activity. This method may be taken as an example of quantitative enzyme cytochemistry, and we can cite its advantages: (i) the activity present in a small element in a cell can be measured selectively; (ii) only the original precipitate is examined, and the later steps for visualization (formation of cobalt sulphide), and their associated possible complications, are eliminated; (iii) the initial activity is readily measured, and therefore the complication due to the deposited precipitate progressively impeding the access of substrate to enzyme can be overcome; (iv) the calcium phosphate can be readily removed by washing, allowing repeated determinations at the same site; (v) by varying the substrate concentration, pH, etc., the kinetics of the enzyme activity can be determined and the enzyme at each site can be characterized by its kinetic characteristics.

Potential developments. In conclusion, we may note some possible developments in this field.

1. The difficulties due to the complex organization of proteins in tissue sections may be turned to account, to give information on this organization. Thus, apparent anomalies or differences in reactivity may allow groups in various kinds of bonding to be detected and measured, e.g. there are indications that we may be able to measure that fraction of arginine which is bound to nucleic acid in nucleohistone.

2. We should expect antibody methods to be improved and exploited on a great variety of proteins. For enzymes that are inactivated by the

preparative procedures, it will be worth while to examine whether the lability of their antigenic activity is less than that of their enzymic activity. The purification of the antigens is a major requirement in this field.

3. The extension of cytochemistry to the electron microscope level would be valuable, although it contains many difficulties. This has already begun, with attempts at the specific deposition of heavy metals.

4. *Isotope methods.* It seems likely that the labelling of reagents with radio-active isotopes will become an important method in future cytochemistry, and I would like to draw attention here to some of its possibilities. A clear distinction must be drawn between metabolic incorporation of isotopic compounds, which has been the subject of a number of cytological studies, and the proposed methods in which a labelled reagent is applied to tissue sections. The techniques of autoradiography are now available for microscopic localization here. The resolution obtained varies with the isotope. C^{14} is satisfactory for some cases, but the greatest advantage will come from using tritium (H^3), since it gives a resolution of 0.5 μ and therefore sharper intracellular localizations, and since it can be readily introduced chemically into a wide variety of organic compounds. Important advantages can be derived from such labelling: (i) An escape is provided from the problems of forming a satisfactory colour at specific groups, since the label can be attached to any suitable simple molecule. (ii) Measurements may be made by grain counting, and this avoids the difficulties due to change of extinction coefficient in molecular aggregates, hyperchromic effects, etc. Useful mutual calibrations of spectrophotometric and autoradiographic measurements of the same group should be possible. (iii) Specific antibodies could be tagged by a small labelled molecule. This would have advantages over fluorescent labelling with respect to certain problems, e.g. measurement, background tissue fluorescence, etc. Some studies with I^{131}-labelled antibody have been made; in view of the poorer resolution and short life of I^{131}, tritium labelling would give further advantage here. (iv) For the localization of enzymes, in one or two cases labelled substrates or trapping reagents have been employed or suggested (13, 14, 15), using P^{32} or radioactive metals. C^{14} or tritium may give advantages here. An alternative approach would be the use of a labelled specific inhibitor, reacting at the active centre of the enzyme. Protection by individual substrates or competitive inhibitors would give finer discriminations.

5. Finally, we may note that we are still far from any approach to microscopic protein cytochemical methods that can be applied to the living cell.

References

1. Mazia, D. "Glutathione: A Symposium." p. 209. S. Colowick *et al.*, eds., Academic Press, New York (1954).
2. Caspersson, T. "Cell Growth and Cell Function." Norton, New York, 1950.
3. Thorell, B. *Cold Spr. Harb. Symp. Quant. Biol.* **12**, 247 (1947).
4. Danielli, J. F. *Cold Spr. Harb. Symp. Quant. Biol.* **14**, 32 (1950).
5. Danielli, J. F. *Symp. Soc. Exp. Biol.* **1**, 101 (1947).
6. Barnard, E. A. and Danielli, J. F. *Nature, Lond.* **178**, 1450 (1956).
7. Holt, S. J., "General Cytochemical Methods." Vol. I, p. 375. J. F. Danielli, ed., Academic Press, New York (1958).
8. Coons, A. H., *Ibid.* p. 399.
9. Barnard E. A. and Stein, W. D. "Advances in Enzymology." Vol. 20, p. 51. F. F. Nord, ed., Interscience Publ., New York (1958).
10. Leuchtenberger, C. This Symposium.
11. Hale, A. J. This Symposium.
12. Barter, R., Danielli, J. F. and Davies, H. G. *Proc. Roy. Soc.* B **144**, 412 (1955).
13. Dalgaard, J. B. *Nature, Lond.* **162**, 811 (1948).
14. Shugar, D., Szenberg, A., and Sierakowska, H. *Exp. Cell. Res.* **13**, 124 (1957).
15. Pearse, A. G. E. *Biochem. J.* **68**, 17P (1958).

CYTOCHEMISTRY OF NUCLEIC ACIDS*

C. LEUCHTENBERGER †

*Institute of Pathology, Western Reserve University
Cleveland, Ohio*

I. INTRODUCTION

CYTOCHEMISTRY, the field which deals with the chemical composition of cells, has interested workers in the biological and medical fields for a long time. Ever since the microscope was discovered and staining methods were developed which allowed the study of the morphology of cells, questions as to chemical components of cells and their behaviour under different conditions have been raised by the cytologist, histologist, and pathologist. But it was actually the biochemist and not the microscopist who gave us the foundation of our present knowledge of the chemical nature of cells.

At the end of the last century two brilliant biochemists, Miescher [37] and Kossel [15], demonstrated for the first time that the essential building stones of all cells of animals, plants and bacteria, are the nucleoproteins, salt-like compounds of nucleic acids and proteins. Their fundamental discovery and the subsequent contributions by other biochemists stimulated the microscopist to relate the biochemical findings to the observations that he made under the microscope. Which structures of the cells contained the nucleoproteins, and how could these compounds be demonstrated *in situ* in the cells or cell parts such as the cytoplasm, nucleus, nucleolus or chromosomes? These were pertinent questions because, in spite of the great contributions resulting from the biochemical studies, such analyses were unable to answer the more specific questions just mentioned. This is mainly due to the fact that the biochemical analysis can be made only after the tissues are macerated and the cells destroyed in order to extract the substances to be

---

* From studies supported in part by research grants A–787, C–1814, and RG–4268 from the National Institutes of Health, U.S. Public Health Service, by a grant from the Brush Foundation of Cleveland, Ohio, and by a grant from the Franchester Fertility Fund of Cleveland, Ohio.

† Present address: The Children's Cancer Research Foundation, Boston 15, Massachusetts.

analyzed. The development of staining procedures for cells, and especially the early effort to interpret the staining of cellular structures in chemical terms, reflect the endeavour of microscopists to correlate structures of cells with chemical constituents. While the use of stains was more or less empirical, this approach actually started an era of what might be called cytochemical research. For review, see Danielli [9], Brachet [5].

Since the work of these early investigators considerable progress has been made in our understanding of the nucleic acids, especially concerning their chemical nature, their intracellular localization and their biological activity in cell-metabolism. The continuous active interest in the nucleic acids and the enormous strides made in this field during the last two decades are well exemplified by the wealth of material that has appeared on this topic in recent publications, reviews, and books (see especially books by Chargaff and Davidson [8] and by Brachet [5]).

Today the fact is well established that there are two types of nucleic acids in the cells, the deoxyribose nucleic acid (DNA) which is characteristic for all nuclei and with very few exceptions is confined to the nuclei and chromosomes *only*, and the ribose nucleic acid (RNA) which is present in the cytoplasm, nucleoli, and nuclei.

The recognition of the biological significance of RNA and DNA has not only led to intensive studies of the nucleic acids in different areas of biology, medicine, chemistry, and physics, but also stimulated the development of many new methods and approaches in this field [6, 41, 10].

II. MICROSPECTROPHOTOMETRY

One of the cytochemical methods that has opened completely new pathways for the study of the nucleic acids, and has been particularly fruitful in elucidating the intracellular location and activities of DNA and RNA is the field of microspectrophotometry. The development of this new type of microscopy has been largely due to the pioneering work of Caspersson in 1936 [6]. Caspersson demonstrated that by extending the optical capacities of the microscope into the analytical sphere, the microscope can be used simultaneously as an instrument for both morphological studies and the chemical analysis of *single* cells or cell parts such as the nucleus, nucleolus, and cytoplasm. Since the analysis can be done *in situ* in microscopic sections, that is without destroying either the architecture of the cells or their relationship within a tissue, *a comparison of cell morphology with chemical composition can be made directly under the microscope.*

Although methodological details cannot be discussed here, it should be mentioned that the basic principle of microspectrophotometry when applied to the determination of the nucleic acids in *single cells* is actually very simple. As in the photometric chemical analyses of solutions, the amount of light absorbed at a specific wave-length by a cell structure is used as a basis for the qualitative and quantitative analyses of the intracellular nucleic acids [6].

There are at present two main microspectrophotometric methods for

Fig. 1. Photograph of microspectrophotometric apparatus as used in the Cytochemical Laboratory at the Institute of Pathology, Western Reserve University. (D) diaphragm lever, (F) filter, (G) galvanometer, (L) light source, (M) microscope, (P) power supply, (PH) phototube housing, (T) telescope.

quantitatively determining the nucleic acids in single cell structures—a direct one and an indirect one. The direct method developed by Caspersson and his school [6, 7] utilizes in unstained microscopic preparations the natural absorption of the nucleic acids in ultraviolet light at 2570 Angstroms, while the indirect one [42] first applies a staining technique specific for the nucleic acids, such as the Feulgen reaction for DNA, and then utilizes the light absorption of the stained structures at a wave length characteristic for the dye nucleic acid complex [20].

Thus microspectrophotometry closely resembles analytical chemistry and a microspectrophotometer such as the one that we use in our laboratories for DNA determinations, and which is shown in Fig. 1, consists actually only of a microscope combined with a device for light absorption measurements [20].

There is the light source L, a monochromator or filters, F, to isolate the desired wave length, the microscope, M, and the photometer head with phototube, PH, and galvanometer, G, which permit the measurement of the amount of light absorbed by a single cell structure. Only light coming from the cell structure to be measured is allowed to reach the phototube; the remainder of the light transmitted is obliterated by a diaphragm (D).

The unique features of this method lie not only in the possibility that a nucleic acid quantity in the order of one trillionth of a gram (10^{-9} mg) can be determined in a well preserved *single* cell structure, such as for example 3×10^{-9} mg DNA in a bull sperm nucleus [31], but that it is also possible to detect differences in quantities of nucleic acid which may occur when a cell undergoes certain changes, such as mitosis, or meiosis, or where variations in amounts of nucleic acids from cell to cell may already exist within a cell population of a tissue. At present, the conventional biochemical methods, in spite of their great contributions to the nucleic acid field, do not permit the study of quantitative nucleic acid changes at the individual cell level because their analyses must be done on relatively large cell populations where individual cell morphology is destroyed and therefore, of necessity, the biochemical technique can yield only average values [33, 29].

III. Results

Among the many problems in which the nucleic acids have been investigated by the special cytochemical technique of microspectrophotometry, only a very few examples can be discussed here. It is hoped, however, that the author is able to convey the main features of the characteristic behaviour of the two nucleic acids in normal and abnormal cells and briefly discuss the role which they play in such important biological processes as growth, heredity, and fertilization.

A. *The nucleic acids in normal cells*

One of the fundamental discoveries which has evolved from the cytochemical and in particular from the quantitative microspectrophotometric studies, is the observation that the intracellular activities of RNA and DNA are strikingly different in normal cells. In spite of the similarity in their chemical configuration, the biological properties of RNA and DNA are nearly opposite. RNA is a very labile and active substance undergoing considerable quantitative changes during cell metabolism and playing an important role during protein synthesis, while DNA is highly stable, its quantity is rather independent of most metabolic processes, and it is closely linked to the chromosomes and the genetic material of the cells.

The direct correlation between RNA and protein synthesis on the one hand and DNA and chromosomal complement on the other, is well exemplified in the case of the hemipteran insect *Arvelius albo-*

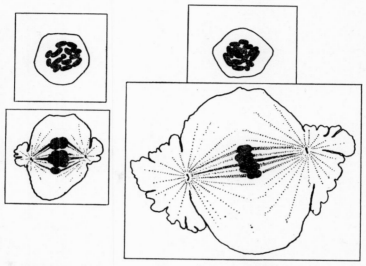

Fig. 2. Chromosomes in normal- and large-sized spermatocytes of *Arvelius albopunctatus*.

punctatus which Schrader and Leuchtenberger studied some years ago by microspectrophotometric techniques [44]. In this insect the primary spermatocytes show differential growth so that the different lobes of the testis contain spermatocytes of strikingly different sizes; for example, the primary spermatocytes in lobe 3 are 4 times the size of those in lobe 2. This difference in size is even more remarkable since chromosomal numbers and the chromosomal size, in other words the chromosomal complement, is the same in both large and normal sized cells (Fig. 2).

When the amounts of DNA, RNA, and proteins were determined in these spermatocytes simultaneously by microspectrophotometry, results were obtained which are summarized in Fig. 3. Taking the actual values for size, DNA, RNA, and protein of the normal-sized cells as 1, it can be seen that the three- to four-fold increase of the cytoplasmic volume in the large-sized cells is parallelled by a similar increase in protein and RNA. The same close link between increase in size, proteins and RNA is found in the nuclei of the large-sized cells, but it can also be noted that, contrary to the behaviour of RNA, the DNA does not change in quantity in the protein synthesizing cells. The constant amount of DNA is, of course, in good agreement with the unchanged chromosomal complement of the large-sized cells.

The *Arvelius* case is just one of the many examples in which not only the parallelism between RNA and protein synthesis can be directly

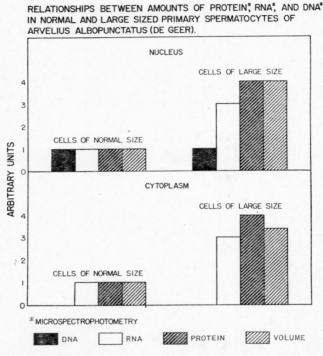

Fig. 3. Relationships between amounts of protein, RNA and DNA in normal- and large-sized primary spermatocytes of *Arvelius albopunctatus* (DeGeer).

demonstrated [7, 5], but it also illustrates that increase in nuclear size must not always be due to increase in chromosomal material, as has been thought to be the case by Hertwig [11].

The unchanged quantity of DNA in these cells and the parallelism with the chromosomal complement are findings that are in good accordance with the DNA constancy theory, first advanced by Boivin, Vendrely and Vendrely [2]. In 1948, these authors, on the basis of their own biochemical studies, reported that the different somatic tissues of a species contain cells with a constant basic amount of DNA which is twice that of the quantity contained in the haploid germ cells of the same species. These observations, and the earlier ones by Avery and co-workers [1] who in 1944 demonstrated that DNA extracted from one strain of bacteria can have a genetic effect upon another strain, gave strong support to the concept that DNA is not only an essential chromosomal constituent, but also an important chemical component of the genetic material. It is not surprising that this idea of harmonious association of chromosomes, DNA, and genes which

RELATIONSHIP BETWEEN CHROMOSOMAL COMPLEMENT AND DNA CONTENT IN SPERMATOGENIC CELLS FROM VARIOUS MAMMALS

TYPE OF CELL	Type of mammals, references and total number of cells (N) measured					
	MOUSE SWIFT 1950 N = 100		BULL LEUCHTENBERGER, SCHRADER ET AL. 1955, 1956 N = 5600		HUMAN LEUCHTENBERGER, SCHRADER ET AL. 1953, 1955, 1956 N = 6000	
	Chromosomal status	Mean amount of DNA per nucleus[1]	Chromosomal status	Mean amount of DNA per nucleus[1]	Chromosomal status	Mean amount of DNA pe nucleus[1]
Primary spermatocyte	4n	4	4n	4	4n	4
Secondary spermatocyte	2n	2	2n	2	2n	2
Spermatid	1n	1	1n	1	1n	1
Sperm	1n	1	1n	1	1n	1

[1] Feulgen microspectrophotometry (arbitrary units)

Fig. 4. Relationship between chromosomal complement and DNA content in spermatogenic cells from various mammals.

fitted so well with Morgan's earlier concept [38] of the genes being located on the chromosomes, greatly stimulated the interest in DNA and gave impetus to studies concerned with the *quantitative* relationship between chromosomal complement and DNA content.

Of great value in the investigation of this special problem have been the microspectrophotometric methods, because they permit the *direct* correlation between amounts of DNA and chromosomal complement from cell to cell on the same preparation.

We have been greatly interested in this particular problem, and in Fig. 4 our own extensive data concerning the relationship between chromosomal complement and DNA content in the spermatogenic cells of bulls and humans are presented [31, 28, 26]. Examining first the DNA values, which are computed mean values based on Feulgen microspectrophotometric determinations on over 11,000 individual spermatogenic nuclei of bulls and humans, it can be seen that if the mean DNA values for the haploid spermatids and spermatozoa are expressed in units of 1 DNA, the DNA values for the primary spermatocytes, secondary spermatocytes, spermatids and spermatozoa are 4 DNA, 2 DNA, and 1 DNA respectively. These ratios of 4 : 2 : 1 of the DNA content are in good accordance with the ratios of the chromosomal status for these cells. The sperms and the spermatids contain the haploid number of chromosomes of 1n; the secondary spermatocytes the diploid number of 2n; and, in the primary spermatocytes, the diploid number of chromosomes but doubling of the chromatids, giving therefore a

RELATIONSHIP BETWEEN CHROMOSOMAL COMPLEMENT AND DNA CONTENT
IN TISSUE CELLS FROM VARIOUS MAMMALS

TYPE OF TISSUE CELL	Type of mammals, references and total number of cells (N) measured					
	MOUSE		BULL		HUMAN	
	LEUCHTENBERGER HELWEG LARSEN ET AL. – 1954, 1956 N = APPROX. 3000 SWIFT – 1950 N = APPROX. 500		LEUCHTENBERGER, VENDRELY, SCHRADER ET AL. – 1951, 1955, 1956 N = APPROX. 2500		LEUCHTENBERGER, ET AL. – 1953, 1954, 1955, 1956 SWARTZ – 1956 HARKIN – 1956 N = APPROX. 5000	
	Chromosomal status	Mean amount of DNA per nucleus[1]	Chromosomal status	Mean amount of DNA per nucleus[1]	Chromosomal status	Mean amount of DNA per nucleus[1]
ADRENAL					2, (4)	2, 4
BREAST					2	2
BRONCHUS	2	2			2	2
COLON					2, (4)	2, (4)
ENDOMETRIUM					2±, 4±	2
KIDNEY	2, (4)	2, (4)	2, (4)	2, (4)	2, (4)	2, (4)
LIVER	2, 4, 8	2, 4, 8	2, 4, 8	2, 4, 8	2, 4, 8	2, 4, 8
LUNG	2, (4)	2, (4)			2	2
LYMPH NODE					2	2
LYMPHOCYTE	2, 4	2, 4			2	2
PANCREAS	2, 4, 8	2, 4, 8	2, 4	2, 4	2, 4, 8	2, 4, 8
PROSTATE					2	2
SEMINAL VESICLE			2	2	2	2
SKIN					2	2
SPLEEN	2	2	2	2	2	2
SPINAL CORD	2	2				
STOMACH					2	2
THYMUS	2, 4	2, 4				
URINARY BLADDER					2, 4	2, 4
SPERMATIDS	1 (Haploid)	1	1 (Haploid)	1	1 (Haploid)	1

[1] Feulgen microspectrophotometry (arbitrary units)

Fig. 5. Relationship between chromosomal complement and DNA content in tissue cells from various mammals.

correspondence to 4n. The DNA data by Swift which he obtained by studying 100 spermatogenic cells of mice show the same relationship [46].

The close association between the chromosomal complement and the DNA content does not only hold for germ cells but also for cells from somatic tissues. In Fig. 5 the relationship between the chromosomal complement and the DNA content in 19 tissues from mice [22], bulls [31, 28], and humans [25, 40] are presented. With the exception of the 500 cells from mice which were analyzed by Swift [46], the DNA analyses were all done in our own laboratories.

Again taking the DNA value of the haploid spermatid as 1 DNA, it can be seen that all the 19 tissues examined contain cells with a basic diploid DNA value of 2. In addition, some tissues, such as liver and pancreas, contain cells showing values of 4 DNA and 8 DNA. It is evident that the chromosomal data correlate very well with the DNA values in all of the tissues except the endometrium. Since, however, there is disagreement among various workers [49, 3] about the chromosomal status of human endometrium, this point will not be discussed here.

The examples of good correlation between chromosomal complement and DNA content given here are from only 3 mammals, but similar findings in other species have been reported [50]. Furthermore, ultraviolet microspectrophotometric studies of the DNA content of the same material are in accordance with these DNA data obtained by Feulgen microspectrophotometry [16].

Although there is indeed an excellent relationship between chromosomal status and the *mean* DNA content of cells, it must be kept in mind that a computed mean DNA value relates to the average cell population, and in order to see whether the good correlation also exists for the DNA content in *individual* cells, the data should be examined from this point of view. In Fig. 6, examples of *individual* DNA values found in human spermatozoa and in human somatic cells are presented. It can be seen that while the mean DNA value for the diploid skin cells is approximately twice that of the haploid sperm, there exists a variability in the DNA content from cell to cell which is more pronounced in the somatic than in the germ cells. The deviations from the expected diploid DNA contents have been usually interpreted as errors inherent in the microspectrophotometric technique [39]. While the variability may in part be due to such technical difficulties, the possibility cannot be excluded that they may be indicative, in part, of biological variations in the DNA content of cells [20].

Support for such a concept comes from recent chromosomal studies by various workers [12, 49, 48] who found that the chromosomal numbers also vary from cell to cell, even in a somatic tissue with a presumed diploid cell population. While most somatic cells carry the basic characteristic diploid chromosome complement corresponding to the characteristic diploid DNA value, there are always cells with chromosomal numbers which are above or below the theoretical value. It is not unreasonable to suggest that cells with variable chromosomal numbers may also have variable DNA values, in other words, the deviations from the expected theoretical value of the chromosomal status are parallelled by similar deviations in the DNA values.

Of special interest is the observation of Tanaka [48] that spermatogenic cells have the least variation in chromosomal numbers, a finding

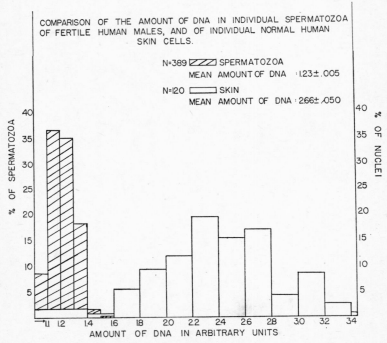

Fig. 6. Comparison of the amount of DNA in individual spermatozoa of fertile human males and of individual normal human skin cells.

which is parallelled by our own DNA results. Comparison of an extensive series of individual DNA data obtained on several thousand normally functional male germ cells with those of somatic cells from the same individuals, always revealed a strikingly lower variability in the nuclear DNA of the germ cells than of the somatic cells.

Although this parallelism between chromosomal and DNA studies supports the view that the variability in some cells may be a truly biological one, the fact remains that most cells of the tissues in a given species are characterized by a basic mean DNA content and a chromosomal complement which Roels [43] designates as an "equilibrium value". While there are some examples in which the basic DNA values change with metabolic and physiological cell activities [45, 29, 20] in the majority of the cases the basic mean DNA value for a given species is very constant in the cells of the various tissues in spite of their metabolic diversities (Fig. 5). There is, however, one cellular activity in which the DNA quantity changes with regularity in the cells, and that is when mitosis takes place. Contrary to what may have been expected from visual inspection of the mitotic process, microspectrophotometry has shown that the period in which the DNA is synthesized, that is when the DNA content is doubled, seems to be relatively brief and does not parallel the period of the complete mitotic cycle. Swift [47] was the first to show that DNA doubling may already be completed in the interphase nucleus.

B. *The nucleic acids in abnormal cells*

The application of microspectrophotometry to the study of nucleic acids in abnormal cells is relatively recent, and systematic comparative studies both of RNA and DNA in abnormal cells have just been started. The work of Caspersson and his school [6, 7] is concerned predominantly with the RNA, while our own studies deal particularly with the DNA. Nevertheless, the available data clearly show that the striking difference that exists between the intracellular activities of RNA and DNA in *normal cells* does not hold regularly in abnormal cells. Although the RNA follows a similar pattern in normal and abnormal cells changing with metabolic activities and parallelling protein synthesis, the DNA deviates strikingly from its stable and constant behaviour as found in normal cells and undergoes considerable quantitative fluctuation in abnormal cells.

During the last eight years we have focused our special attention on the DNA in abnormal cells, and we have studied a variety of pathological conditions [19, 24]. It was felt that the unique opportunity afforded by microspectrophotometry for carrying out simultaneous chemical and morphological analyses on the same cell structure might advance our knowledge concerning the role of the DNA for etiology and pathogenesis, and might help greatly in the elucidation of the sequence of events leading to disease. Furthermore, it was hoped that the early diagnosis of disease would be greatly aided since it was found that microspectrophotometry also offers the possibility of detecting changes in DNA before structural alterations in cells manifest themselves under the microscope [26]. Furthermore, the stability and constancy of DNA in normal cells seemed to be an excellent basis for comparison with the DNA behaviour in cells from an organism that undergoes a pathological process.

One of the aspects that we investigated dealt with the following questions: Is the DNA content in tumour cells the same as in normal cells? Is there a similar concurrence between chromosomal status and DNA content as found in normal cells? *A priori*, it was reasonable to expect that the DNA content of tumour cells would be larger than that in normal cells. This concept was suggested by the increase in size and staining density and by the mitotic abnormalities so frequently observed in tumours. Since very little was known about the DNA content of cells in *normal human* tissues prior to our own studies [25, 40], a comparative extensive study on the DNA content of a variety of normal and malignant human tissues had to be carried out to establish a base line. DNA measurements were made of nearly 10,000 individual cells from 49 normal and 28 human malignant tissues. As discussed before and as seen in Fig. 5, it was found that all the normal human tissues, regardless of their origin or metabolic function, contained cells with a similar constant basic mean DNA content corresponding to the diploid chromosomal complement.

In contrast to the constant and orderly pattern of DNA in cells of normal tissues, the DNA content of precancerous lesions and malignant tumours was considerably higher, and revealed a much larger variability from cell to cell than the DNA in cells of normal homologous tissues. An example of the DNA values of cells from a malignant stomach tumour with metastasis to lymph nodes is given in Fig. 7. It can be seen that the DNA values in the cells of the primary stomach tumour and in the cells from the malignant lymph nodes show higher values and larger variability than the DNA values in the cells of the corresponding normal tissues.

While deviating and higher DNA values were found in all of the malignant human tumours that we have studied so far, they cannot be considered a specific criterion for malignant transformation of cells, but may be explained mainly on the basis of growth and mitotic processes present in most tumour tissues. As pointed out before, DNA synthesis occurs also in cells of normal tissues which are undergoing mitosis and growth. On the other hand, since most normal tissues in adult humans do not exhibit mitoses but show the usual characteristic DNA value with relatively little variability, an increase and a large scatter of DNA in cells of such a tissue must be looked upon with suspicion in regard to malignancy, unless regeneration is to be expected.

The direct relationship between chromosomal complement and DNA values as found in normal cells does not seem to hold for all tumour cells. Although our own work on mouse ascites tumours [23, 17, 18] showed that a diploid tumour had a basic value of 2 DNA and a tetraploid tumour a basic value of 4 DNA, our studies on a large variety of human tumours do not show this parallelism in every case. For example, Koller [14] reported that the diploid number of chromosomes was characteristic for nearly all human tumours. Consequently the basic DNA value in human tumours should be 2 DNA corresponding to the diploid number of chromosomes and because of mitosis, value between 2 and 4 DNA should be obtained. However, very often we actually found basic DNA values of 4 DNA and higher, indicating that multiples of DNA values in human tumours are not necessarily

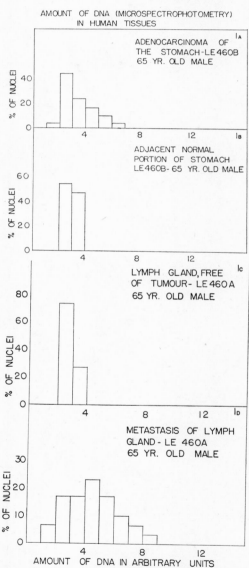

Fig. 7. Amount of DNA (microspectrophotometry) in human tissues.

synonymous with polyploidy, but may be due either to polyteny or to an increase in DNA without a change in chromosomal status [18].

Another example in which the DNA shows a striking increase in amount is illustrated in cells infected with virus. Because of the chemical composition of viruses (many of them contain DNA) and their peculiar relationship to cells, measurements of DNA by microspectrophotometry seemed to afford intriguing possibilities for disclosing the presence of viruses in cells. In virus diseases such as *molluscum contagio-*

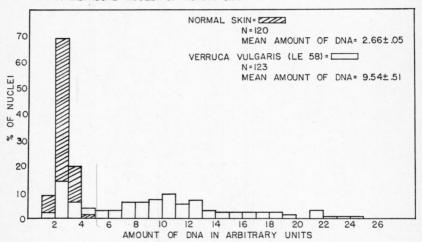

Fig. 8. Amount of DNA (microspectrophotometry of Feulgen reaction) in individual nuclei of human skin.

sum or *verruca vulgaris*, multiplication of the viruses within the infected cells has been demonstrated [7]. If Feulgen microspectrophotometric studies of DNA are done on such virus-infected cells, unusually large quantities of DNA can be found in individual cells. A characteristic example is shown in Fig. 8. It is evident that the greatly increased quantity of DNA in human skin cells infected with the common wart virus, verruca vulgaris, is in marked contrast to the 2 DNA content of normal skin cells. The question arises as to whether this striking increase in DNA is an expression of virus multiplication or whether it reflects chromosomal behaviour or possibly both. Since there are no mitoses present which would account, at least in part, for the larger DNA values, and since there is a corresponding increase of nuclear sizes of these virus-infected cells, it may be suggested that the DNA increase is associated with the formation of polyploidy or polyteny. As a matter of fact, when we first studied the DNA content of virus-infected cells we assumed that the increase of DNA was correlated with the formation of polyploidy or polyteny [36]. However, careful cytological analysis of virus-infected cells did not support such a concept, but instead revealed an altogether different picture from that of polyploid or polytene cells. A good illustration of the virus host cell relationship is given in a recent analysis by Boyer, Leuchtenberger and Ginsberg [4] on tissue cultures of HeLa cells infected with viruses of the upper respiratory tract of man. Such material is especially favourable because

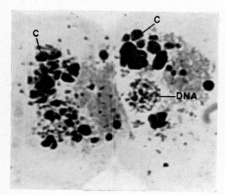 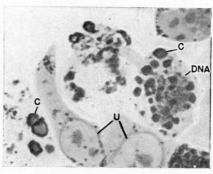

Fig. 9a (left) and 9b. HeLa cells 63 hours after infection with AD III virus, fresh preparation unfixed and unstained. Appearance under the interference microscope (Baker 40 X) magnification approx. 1600. (C) DNA-containing crystals, (DNA) intranuclear DNA masses, (u) uninfected HeLa cells. Note differences in size of nuclei of uninfected cells and virus-infected cells.

it permits simultaneous studies of the sequential changes in the cytology and the DNA content at different time periods after virus infection. The occurrence of well defined inclusion bodies, the increase in DNA and nuclear sizes, the accumulation of DNA-containing masses and the presence of DNA-containing crystals in the nuclei of the virus-infected cells [21] are findings consistent with the presence of intranuclear virus and are completely different from those found in polyploidy or polyteny. The appearance of the DNA-containing masses and crystals in the nuclei of virus-infected cells is shown in Fig. 9.

Our own studies concerning the problems of male infertility [35] and of dwarfism [31, 30] clearly revealed that abnormal processes are not always associated with an increase in DNA as in tumours and virus diseases.

In contrast to the remarkably constant and uniform haploid DNA values found in the spermatozoa of fertile males (both human and bulls), the DNA content of the spermatozoa from infertile males was shown to be variable and significantly lower [32, 34]. This finding is especially pertinent since the cytological appearance and the dry mass values of spermatozoa containing normal and deficient amounts of DNA are identical [28]. The same holds true for the spermatogenic cells; primary and secondary spermatocytes and spermatids each have a deficient DNA content in infertile males when compared to the 4 DNA, 2 DNA, and 1 DNA found in the fertile males, although the histological and cytological features of the testes and germ cells may be completely normal [26]. It thus appears that the DNA deficiency found

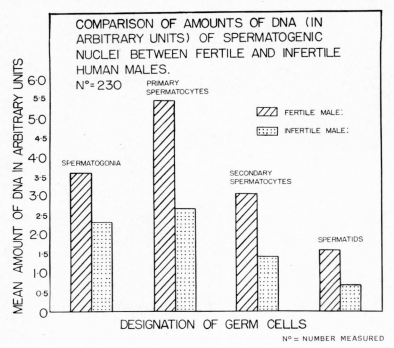

Fig. 10. Comparison of amounts of DNA (in arbitrary units) of spermatogenic nuclei between fertile and infertile human males.

by Feulgen microspectrophotometry in spermatogenic cells of normal cytological appearance can be considered a criterion for gauging at least one type of male infertility and may have a bearing on the fertilization process [13].

Similar DNA deficiencies in the spermatogenic tissue and spermatozoa were also found in our studies of dwarf bulls, and in one case of a sterile dwarf bull, an extreme DNA deficiency was even found in both somatic and spermatogenic cells.

While, as pointed out previously, only a very few examples in which the nucleic acids have been studied were discussed here, it is hoped that this brief assessment of results may help in our attempt to understand the roles of nucleic acids in the living organism. There are many unexplored fields in biology and pathology where microspectrophotometry may prove of great value especially when the resulting data are correlated closely with other experimental methods on the same material.

References

1. AVERY, O. T., MACLEOD, C. M. and McCARTY, M. *J. Exp. Med.* **79**, 137 (1944).
2. BOIVIN, A., VENDRELY, R. and VENDRELY, C. *C.R. Acad. Sci, Paris.* **226**, 1061 (1948).
3. BOOTHROYD, E. R. and WALKER, B. E. *Genetics* **37**, 567 (1952).
4. BOYER, G. S., LEUCHTENBERGER, C., and GINSBERG, H. S. *J. Exp. Med.* **105**, 195 (1957).
5. BRACHET, J., "Biochemical Cytology." Academic Press Inc., New York (1957).
6. CASPERSSON, T. *Skand. Arch. Physiol.* **73**, suppl. 8, 1 (1936).
7. CASPERSSON, T. "Cell Growth and Function." W. W. Norton and Co., New York (1950).
8. CHARGAFF, E. and DAVIDSON, J. N. "The Nucleic Acids." Vol. I and II. Academic Press Inc., New York (1955).
9. DANIELLI, J. F. "Cytochemistry. A Critical Approach." John Wiley & Sons, Inc., New York (1953).
10. DAVIES, H. G., WILKINS, M. H. F. and BODDY, R. G. H. B. *Exp. Cell Res.* **6**, 550 (1954).
11. HERTWIG, G. *Z. Mikr.-Anat. Forsch.* **45**, 37-45 (1939).
12. HSU, T. C. and POMERAT, C .M. *J. Morphol.* **93**, 301 (1953).
13. ITO, S. and LEUCHTENBERGER, C. *Chromosoma* **7**, 328 (1955).
14. KOLLER, I. *Ann. N. Y. Acad. Sci.* **63**, 793-817 (1956).
15. KOSSEL, A. *Arch. Anat. Physiol., Lpz.* (Physiol. Abt.), 359 (1891).
16. LEUCHTENBERGER, C. *Science* **120**, 1022 (1954).
17. LEUCHTENBERGER, C. *Exp. Cell Res.* **11**, 506 (1956).
18. LEUCHTENBERGER, C. *Ann. N. Y. Acad. Sci.* **63**, Art. 5, 816 (1956).
19. LEUCHTENBERGER, C. *J. Mt Sinai Hosp.* **24**, 971 (1957).
20. LEUCHTENBERGER, C. "Quantitative Determination of DNA in Cells by Feulgen Microspectrophotometry, in General Cytochemical Methods." Vol. 1, Academic Press, Inc., New York (1958).
21. LEUCHTENBERGER, C. and BOYER, G. *J. Biophys. Biochem. Cytol.* **3**, 323 (1957).
22. LEUCHTENBERGER, C., HELWEG-LARSEN, H. F. and MURMANIS, L. *Lab. Invest.* **3**, 245 (1954).
23. LEUCHTENBERGER, C., KLEIN, G. and KLEIN, E. *Cancer Res.* **12**, 480 (1952).
24. LEUCHTENBERGER, C. and LEUCHTENBERGER, R. *Schweiz. Med. Wschr.* **87**, 1549 (1957).
25. LEUCHTENBERGER, C., LEUCHTENBERGER, R. and DAVIS, A. M. *Amer. J. Path.* **30**, 65 (1954).
26. LEUCHTENBERGER, C., LEUCHTENBERGER, R., SCHRADER, F. and WEIR, D. R. *Lab. Invest.* **5**, No. 5, 422 (1956).
27. LEUCHTENBERGER, C., LEUCHTENBERGER, R., VENDRELY, C. and VENDRELY, R. *Exp. Cell Res.* **3**, 240 (1952).
28. LEUCHTENBERGER, C., MURMANIS, I., MURMANIS, L., ITO, S. and WEIR, D. R. *Chromosoma* **8**, 73 (1956).

29. LEUCHTENBERGER, C. and SCHRADER, F. *Proc. Nat. Acad. Sci., Wash.* **38**, 99 (1952).
30. LEUCHTENBERGER, C. and SCHRADER, F. *J. Biophys. Biochem. Cytol.* **1**, 6. 615 (1955).
31. LEUCHTENBERGER, C., SCHRADER, F., HUGHES-SCHRADER, S. and GREGORY, P. W. *J. Morph.* **99**, 481-512 (1956).
32. LEUCHTENBERGER, C., SCHRADER, F., WEIR, D. R. and GENTILE, D. P. *Chromosoma* **6**, 61 (1953).
33. LEUCHTENBERGER, C., VENDRELY, R. and VENDRELY, C. *Proc. Nat. Acad. Sci., Wash.* **37**, 33 (1951).
34. LEUCHTENBERGER, C., WEIR, D. R., SCHRADER, F. and MURMANIS, L. *J. Lab. Clin. Med.* **45**, 851 (1955).
35. LEUCHTENBERGER, C., WEIR, D. R., SCHRADER, F. and LEUCHTENBERGER, R. *Acta Genet.* **6**, 272 (1956).
36. LUND, H. and LEUCHTENBERGER, C. *Cancer Res.* **12**, 278 (1952).
37. MIESCHER, F., "Histochemischen und Physiologischen Arbeiten." Vogel, Leipzig (1897).
38. MORGAN, T. H. "The Genetic and the Operative Evidence Relating to Secondary Sexual Characters." Carnegie Inst., Washington, 1919.
39. PATAU, K. and SWIFT, H. *Chromosoma* **6**, 149 (1953).
40. PERSKEY, L. and LEUCHTENBERGER, C. *J. Urol.* **78**, 788 (1957).
41. POLLISTER, A. W. *Rev. Hémat.* **5**, 527 (1950).
42. POLLISTER, A. W. and RIS, H. *Cold Spr. Harb. Symp. Quant. Biol.* **12**, 147 (1947).
43. ROELS, H. *Nature, Lond.* **174**, 514 (1954).
44. SCHRADER, F. and LEUCHTENBERGER, C. *Exp. Cell Res.* **1**, 421 (1950).
45. SCHRADER, F. and LEUCHTENBERGER, C. *Exp. Cell Res.* **3**, 136 (1952).
46. SWIFT, H. *Physiol. Zoöl.* **23**, 169 (1950).
47. SWIFT, H. *Proc. Nat. Acad. Sci., Wash.* **36**, 643 (1950).
48. TANAKA, T. *Kromosomo* **2**, 39 (1951).
49. TIMONEN, S. *Acta Obstet. Gynec. Scand.* Supp. 2, 1 (1950).
50. VENDRELY, R. and VENDRELY, C. "L'Acide Désoxyribonucléique." Amédée le Grand & Cy, Editeurs (1957).

THE INTERFERENCE MICROSCOPE AS A CELL BALANCE

A. J. HALE[*]

Institute of Physiology, University of Glasgow

INTRODUCTION

THE interference microscope is no innovation, the first having been designed in 1894 by Sirks, [30] but its usefulness as a practical tool in cytological research has only recently become apparent [5, 15].

Many different designs of microscope have appeared but due to mechanical and optical limitations in the design, these instruments could not be produced commercially and thus were not readily available to cytologists.

When the designs of Dyson [10], Smith [31] and Johansson [21] appeared and were manufactured, interference microscopes became available at a cost little more than that of a conventional light microscope.

The great value of the interference microscope is that

(a) It permits observation of living or fixed cells in constantly variable colour contrast and without the sometimes undesirable halo effects of phase contrast microscopy.

(b) It permits accurate measurement of the optical thickness of a microscopic object.

(c) Since this optical thickness is a function of refractive index and geometric thickness, it can be used to compute one of these components if the other is known.

(d) As the refractive index is related to the concentration of material within the object, then the concentration and also the total mass of the object can be calculated.

It thus provides at reasonable cost and in a relatively simple way a method of making accurate absolute measurements of some of the fundamental properties of cells.

---

[*] Present address: Division of Pathology, Imperial Cancer Research Fund, Lincoln's Inn Fields, London, W. C. 2.

Principle of Interferometry

When light passes through a hole or a slit in a piece of cardboard it is diffracted out into the area around the slit in a regular manner. If the waves of light from this slit are allowed to pass through two similar slits a short distance away from the original then two similar wave fronts will emerge from these slits (Fig. 1). These two wave fronts will overlap

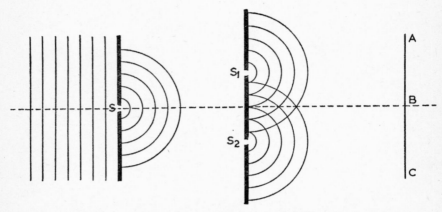

Fig. 1. The pattern of advancing wave fronts passing through one and then two slits. The wave fronts are in phase as they are derived from a common source.

and where they do so will interfere with one another. Interference is the phenomenon which occurs when two wave trains interact so that the amplitude of each train augments or destroys the amplitude of the other. This will be seen in Fig. 2 where the peak of one train is nearly opposite the trough of another. These opposing deflections produce a

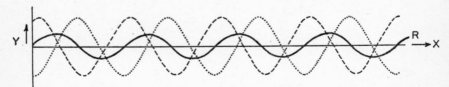

Fig. 2. Superposition of two wave trains of the same frequency and amplitude but markedly out of phase. The wave train R represents the resultant which has the same frequency but a smaller amplitude.

resultant wave form which has an amplitude less than either of the initial waves, but has the same wavelength (destructive interference). If two peaks are nearly superimposed then the resultant wave will have an amplitude greater than either of the two interfering waves, but again will have the same wavelength (constructive interference) (Fig. 3). If a ground-glass screen is put some distance in front of the two slits, as in Fig. 1, then a series of dark and light bands would be seen (Fig. 4). The dark bands represent points of destructive interference and the light bands the points of constructive interference. If

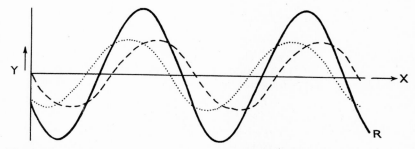

Fig. 3. Superposition of two wave trains of the same frequency and amplitude but which are slightly out of phase. The wave train R represents the resultant which has the same frequency but a greater amplitude.

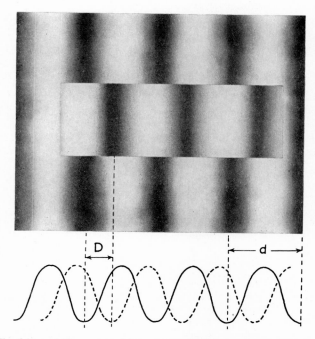

Fig. 4. The fringe shift produced in a fringe field by a large object with an even optical thickness across its surface. The broken curve indicates the variation in intensity of the fringe pattern in the object.

white light (light of mixed wavelengths) is used then the bands will be coloured. The colour pattern will depend on the spectral output of the light source and the way in which interference is produced.

This is the basic phenomenon of interference which occurs irrespective of how it is produced.

DESIGN OF INTERFERENCE MICROSCOPES

Of all the designs of interference microscope that have been produced the ones of greatest practical value are those of Dyson [10], Smith [31] and Johansson [21] which are manufactured by Cooke, Troughton and Simms, C. Baker of Holborn and Jugnarbolaget of Stockholm, respectively. Although each of these microscopes utilizes the same basic principle of interference in its design, the way in which interference is produced is different in each one. This is so for a definite reason. All interference microscopes have some limitation in their design [11]. The numerical aperture may be restricted, the use of immersion objectives may be required or there may be more than one image in the field. Because of this the designer must decide which restrictions he can tolerate and which he must dispense with. Thus one microscope may be more suitable for examination of a particular type of specimen than another. I have listed the advantages and disadvantages of the available microscopes in Table 1.

TABLE 1

Microscope	Advantages	Disadvantages
Cooke-Dyson	(1) Single image in the field. (2) Freedom from haloes. (3) Constantly variable interference. (4) High numerical aperture.	(1) Immersion objectives must be used. (2) Very sensitive to adjustment.
Baker-Smith	(1) Dry objectives can be used. (2) Easy to adjust.	(1) Two images in the field. (2) Interference is not constantly variable.
Jugnarbolaget-Johansson	(1) Free from error in measurement caused by numerical aperture. (2) Easy to adjust.	(1) Two images in the field. (2) Numerical aperture is restricted. (3) Requires accurate calibration. (4) Interference is not constantly variable.

It is advisable to give due consideration to these points before purchasing a microscope for a particular purpose.

Numerous pieces of ancillary equipment can be bought for use in making measurements with these microscopes. Some of these can be supplied by the microscope manufacturers.

Interpretation of the Image

In the interference microscope it is possible to manipulate the components so that the microscope field may or may not contain interference bands. The bands in the field can be made very close together (narrow) or very far apart (broad). If the broadening is carried out to a marked extent (tending to infinity) then the field will appear to be evenly illuminated across its diameter. The colour of this field will depend upon the light source and the band in the system selected. By changing the band being examined the colour of the field will alter. Some form of calibration is required so that the distance, in units of measurement relative to the wavelength of light, can be measured between adjacent bands in the interference pattern.

When a specimen is placed in an interference field it will have an intensity different from the background. The degree of difference will depend on the optical thickness (refractive index × geometric thickness) of the specimen and its position in the bands of the field.

When light passes through a specimen it will in general be slowed down by an amount dependent on the optical thickness. Interference is produced by two sets of wave fronts. One set can be considered as passing through the specimen (specimen beam) and the other as passing around it (comparison beam). The specimen beam will thus be slowed down relative to the comparison beam. This means that in the specimen beam the peak of a wave will lag behind the peak in the comparison beam. The effect of this is that the interference pattern in the image of the specimen will be different from that of the background. For example, if a specimen is examined in a field showing maximum constructive interference (bright field) and it has slowed one beam by one half wavelength, then it will show destructive interference and will appear dark. The degree of contrast between the specimen and the field will depend on its position in the field. For example, if the specimen described above was examined in a dark field it would appear bright.

If a long narrow strip of glass of uniform thickness is immersed in water and examined in the microscope, then the bands in the background will be equally displaced at all points on the strip (Fig. 4). The amount of this displacement is a measure of the optical thickness. If the strip is thinner at one end, the displacement of the bands will be less at the thin end than at the thick. If several small pieces of glass equivalent in thickness to the thin, the intermediate, and the thick ends of the wedge were examined, then each piece would displace the bands by the same amount as the point in the wedge of the same thickness. As the pieces are small, however, one does not see a displacement of a band but merely a difference in intensity, between the object and the background, which is equivalent to the band displacement seen in the wedge of glass. Thus the same object will be dark or light depending upon its position in the field. In white light the contrast will be coloured; each band being of a different hue.

If an object has a small projected area but a large optical thickness e.g. a glass bead, and if it slows the light by several wavelengths, a

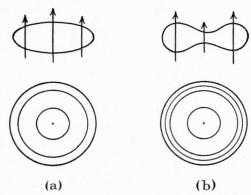

Fig. 5. Objects of the shape shown in the upper part of the diagram will have the appearance shown in the lower diagrams if their optical thickness is greater than one wavelength.

pattern similar to Newton's rings will be seen in the object. This and other explanations can be given for simple test objects, but cells are normally more complicated than these.

The optical thickness of an object is seldom even across its surface, due either to variations in refractive index, geometrical thickness or both. Consequently it may be difficult to interpret a pattern. e.g. a sphere could give a pattern similar to that of a biconcave disc (Fig. 5). It is possible, however, to detect differences in these similar images by correct use of the microscope.

The contrast in an interference system can be explained according to wave theory [2]. The interpretation of the contrast of many small objects is, however, often difficult, and a full consideration of the theory and practice of interference microscopy is necessary before any interpretations are attempted. The image obtained is often markedly different from that seen by phase microscopy where contrast may be artificially enhanced in several ways.

Methods of Measurement of Optical Thickness

The distance between two dark bands, or two light ones, in an interference field is one wavelength (λ) irrespective of whether the bands are 1 cm or 10 cm apart. The value of λ is taken from the wavelength of light used (e.g. 5461 Å in mercury green light). If white light is used, then the distance between adjacent coloured bands is approximately one wavelength. If a simple object, like the strip of glass mentioned above, is examined in a coloured field filled with bands, then the amount of slowing of the light by the object (optical retardation) can be approximately measured by counting (in number of bands) the displacement of bands of similar hue between the bands in the object and the bands in the background. This will give you the number of wavelengths of optical retardation e.g. $2n = 2\lambda = 2 \times 5461$ Å. This is only an approximation to one wavelength. Similarly, if the object is a sphere and it is examined in a field where the bands have been broadened, so that the field is filled by one part of a band, then the object contrast will appear to be a series of Newton's rings superimposed on a field of one colour. Counting the number of rings will give a rough measure of the optical retardation.

Various improvements on this technique have been devised and described in detail [5, 15]. Using these improved methods it is possible with certain objects and good conditions of examination to measure the optical retardation with an accuracy of $1/300\lambda$, i.e. to $5461/300$ Å $= 18.0$ Å.

The choice of a particular method of measurement and the accuracy attainable will depend on
(a) the size of the object.
(b) whether or not it has an even optical thickness.
(c) whether it is discrete or in close association with other objects.
(d) the motility of the object.
(e) the type of microscope available.

The Cytochemical Significance of Measurements

As we have seen, an object retards light because of the difference in optical path between it and the surrounding medium. The amount of the retardation depends upon the refractive index (n) and the thickness (t) of the object. Water has a refractive index of 1.333 and living cells have a refractive index varying from 1.354 to 1.365. Fixed cells usually have a refractive index of about 1.54.

The optical retardation (Δ) of the object is

$\Delta = n.t$ if it is mounted in air or
$\Delta = (n_o - n_m) t$ when the object of

refractive index n_o is mounted in a medium of refractive index n_m.

As Δ can be measured accurately in the interference microscope, then if either n_o or t can be measured by some means then the remaining component can be measured

$$\text{e.g. } t = \frac{\Delta}{(n_o - n_m)}$$

thus if n_o, n_m and Δ are measured, t can be calculated.

$$\text{Now } \alpha = \frac{n_o - n_w}{C}$$

where α is the specific refractive increment, n_o and n_w are the refractive indices of a solution (the cell cytoplasm) and water respectively and C is the concentration in g/100 ml of solution.

In most biological materials α is nearly constant (~ 0.0018); therefore, by measuring the refractive index of the object one measures the concentration of material within it. Similarly, it can be shown that the total dry mass (m) (weight of material other than water) of a cell is related to Δ in the following way.

$$m = \frac{\Delta \cdot A}{\chi}$$

where A is the area of the object in cm^2 and χ is 100α. Thus by measuring Δ in the interference microscope and A by planimetry, m can be calculated since χ is a constant (0.18).

Various correction factors have to be applied to the equation depending upon whether the cell is living or fixed and depending upon the refractive index and certain other characteristics of the medium in which the cell is mounted.

To obtain the maximum amount of information about a cell or cell component, one must measure Δ when the object is examined consecutively in two media of different refractive indices. This is a delicate technique but the results are rewarding. Using the two measures of Δ one can calculate

(a) the total dry mass.
(b) the refractive index of the object and thus the concentration of material within it.
(c) the thickness of the object.
(d) the volume.
(e) the wet mass of the object.

The accuracy of these measurements will, of course, depend upon the factors influencing the accuracy of measurement of Δ. Under good conditions the dry mass can be determined with an accuracy of 0.5×10^{-12}g, the concentration with an accuracy of $\pm 1.0\%$ and the thickness to $0.05\,\mu$ (500 Å).

Sources of Error

As the interpretation of the optical retardation in terms of absolute values or cell constitution depends upon the constancy of α, it should be pointed out that α is not constant for all biological materials. It is fortunate, however, that deviations from constancy only occur with fat and inorganic salts when these are present in relatively pure form. When these substances are combined with proteins and other cell constituents then the deviation from a value of 0.0018 is small.

Errors in measurement may be produced by the microscope system in several ways. These are by

(a) Glare.
(b) Light scatter.
(c) Absorption.
(d) Phase changes in the comparison beam.
(e) Existence of a finite condenser aperture.

All of these may lead to an error, the magnitude of which will vary with the size of the object and its physical characteristics and the refractive index of the medium in which it is mounted. With a specimen of reasonably simple optical characteristics and an accurate objective method of measuring the optical retardation, the error in measurement should not be greater than 10%.

Validity of the Method

The only way in which it has been possible to assess the validity of the method is to compare measurements made by it with those obtained by means such as bulk biochemical analysis, ultra-violet microspectrophotometry and historadiography.

The ratio of dry weight to deoxyribonucleic acid in ram sperm is similar when measured by interferometry and microspectrophotometry to the ratio obtained by biochemical analysis [9]. The dry weight of nuclei isolated from the thymus of the calf is the same when measured by interferometry and by biochemical analysis [17]. The results of a comparison of measurements, of mass per unit area in histological sections of several tissues, made by interferometry and historadiography show a close correlation [8].

Biological Value of Interference Microscopy

At first sight the interference microscope may appear to many to be an interesting physical toy of limited application to biological problems. Consideration will show that it is an accurate and valuable instrument, which might be called the balance of the microscopist.

So used, by a microscopist who realizes its potentialities and its limitations, it proves an extremely valuable instrument in cytological research.

Apart from the quantitative measurements which can be made, the interference microscope can be used for investigating the morphology of cells and tissues [1, 10, 29] (Fig. 6). Huxley [18, 19, 20] has made use of his design of microscope to investigate the morphological changes in the sarcomeres of living muscle during contraction and has made a beautiful ciné-film of these changes. The images seen in the interference microscope are free from the haloes which are seen with the phase contrast microscope. These haloes are usually the result of abrupt changes in refractive index within the object. Because of the absence of haloes certain specimens may appear rather featureless in the interference system. In addition as the contrast of any part of the specimen varies with its optical retardation and its position in the field and as two parts of the specimen may have different optical retardations, then maximal contrast of these two parts will not be obtained at the same position in the field. Thus the specimen must be examined in different positions in the field to obtain good images of these two parts. Both of these effects may produce contrast in a specimen, or part of it, which is not as great as one would get with the same specimen using a phase contrast microscope. The images one does see are, however, a more accurate representation of the object.

The mass values of the nuclei of fibroblasts cultured from the heart of chicken embryos have been found [13] to range from 9 to 40 picograms (g $\times$ 10^{-12}). These observations on the nuclei of intact cells may not be absolutely correct, since there is presumably a thin layer of cytoplasm above and below the nuclei in these cells [3]. The average dry mass value of the nuclei of mouse fibroblasts has been found to be 74 picograms compared with an average of 88 in the nuclei of sarcoma cells [27]. A linear relationship has been found between the dry mass of protoplasm and the area of the cell whether it is a fibroblast or a sarcoma cell. The dry mass per unit area of cytoplasm was approximately 40.0 picograms/μ^2 in fibroblasts and 44.0 in sarcoma cells. This corresponds to a range of 3.4 to 24.9 picograms/μ^2 in the cytoplasm of chick fibroblasts [3]. This latter investigation demonstrated the very interesting point that the concentration of material is constant throughout the cytoplasm of any cell at one period of time, and also that the dry mass of the cell is independent of changes in the osmotic pressure of the medium in which it is mounted. It has also been shown [17] that if nuclei are separated from cytoplasm in aqueous media, then there is a considerable loss of protein from the nuclei. Those nuclei isolated in non-aqueous media have a dry mass of 35 picograms, whereas those isolated in aqueous media may have as little as 15 pico-

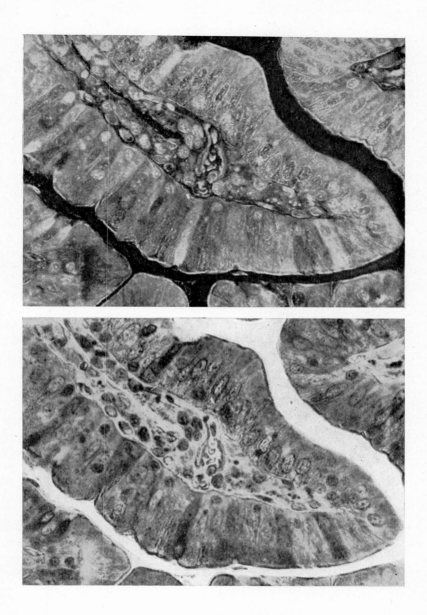

Fig. 6. Photographs of a 4 μ section of small intestine taken at infinite fringe separation in the Baker system. The upper picture shows destructive interference in the background whereas the background shows constructive interference in the lower photograph.

grams left. The thickness of the cytoplasm of fibroblasts of chick [3] and mouse and of sarcoma cells of the mouse [27] has been measured, but as considerable variation may be found within any one cell and because the cell may swell when mounted in a hypotonic medium, these measurements are of little value for comparative purposes.

The dry mass of sperm heads has been found to be 16 picograms in the mouse when the cells are unfixed and 13 when fixed [28]. Similar values for the rat are 11.8 and 11.1 respectively, and in the guinea-pig the value is 10.9 for fixed sperms [26]. The value for fixed sperm from fertile bulls was found to be 7.1 picograms compared with 7.3 in infertile bulls [23]. The dry weight of fixed ram sperm is 7.2 picograms [9]. The values for mouse, rat and guinea-pig were obtained assuming that the refractive index of fixed sperm heads is 1.59. This is not a safe assumption, since it has been shown that the refractive index of fixed ram sperm is 1.54 [9]. Because of this assumption the formula used by these investigators in the calculation of their results introduces a considerable error in their computations.

An interesting feature of the relative constancy of the dry weight of the sperm heads in fertile and infertile bulls is that the arginine contents of these two groups of sperms were the same but their deoxyribonucleic contents were different.

It has been reported that the dry mass of normal red blood corpuscles from one person varies from about 15 to 45 picograms with a mean of 29, whereas in one case of pernicious anaemia the dry mass per cell varies from 15 to 73 picograms with a mean at 38 [25]. I have found [12] that there is a great variation in optical retardation across the surface of a normally biconcave corpuscle, and that some method of integration must be used if an accurate measurement of dry mass is to be made. There is a very small variation in the cells of any one individual. There is a variation between individuals which shows that as the whole blood haemoglobin falls, within physiological limits, the optical retardation at the centre of the corpuscle increases, while that towards the periphery drops slightly. This appears to be an extension into normality of the phenomenon of tendency to spherocytosis which occurs in anaemias. It has also been found [16] that if one measures the refractive index (n_o) of a homogeneate of packed corpuscles, the mean corpuscular volume of a sample, the average retardation (Δ) of the corpuscles and their average projected area (A), and the refractive index (n_p) of the plasma in which they are mounted, then

$$\Delta \times A = (n_o - n_p) \text{MCV}$$

which is what one would expect from theory.

In the early stages of development of erythroblasts, there is an increase in cytoplasmic dry mass, whereas in the later stages of development the mass remains constant as this accumulated material is converted to haemoglobin [22]. The interference microscope has also been used to measure the rate of haemolysis of red blood corpuscles [24].

Using interference microscopy and historadiography, it has been shown that the organic fraction (collagen) in compact bone is relatively constant throughout the bone irrespective of the presence or absence of Haversian systems. On the other hand, the inorganic material (hydroxyapatite) is distributed in the Haversian canals in a definite pattern [7].

The interference microscope can be used to make comparisons of the mass per unit area in different parts of the same histological section, and by this means it has been shown that certain cancer cells have a much higher mass than normal cells [6]. It is also possible to make comparisons between different histological sections if the experiment is properly designed [14].

A most interesting application of the technique is in the investigation of the amount of calcium deposited at sites of enzyme activity during the Gomori-Takamatsu method for alkaline phosphatase. Using this approach much of the basic kinetics of the reaction can be investigated [4].

There is no doubt of the validity and accuracy of observations made using the interference microscope. There is a great temptation to measure dry mass etc. with an accuracy that is greater than is justified. In comparable measurements made by microspectrophotometry it is only comparatively recently that it was accepted that differences of less than 100% were of any biological significance. The accuracy in interference microscopy is of the order of 10%, and when the technique is used in conjunction with other quantitative cytochemical methods a considerable amount of detailed and accurate information can be obtained about single cells.

Acknowledgments

I wish to thank Messrs. E. & S. Livingstone of Edinburgh for permission to reproduce Figs. 1, 2, 3, and 4 from my monograph on the interference microscope.

References

1. Andrews, A. M. and Hale, A. J. *Lab. Invest.* **3**, 58 (1954).
2. Barer, R. *J. R. Micr. Soc.* **75**, 23 (1955).
3. Barer, R. and Dick, D. A. T. *Exp. Cell. Res.* Suppl. **4**, 103 (1957).
4. Barter, R., Danielli, J. F. and Davies, H. G. *Proc. Roy. Soc.* B **144**, 412 (1955).
5. Davies, H. G. "General Cytochemical Methods." Vol. 1, Academic Press Inc., New York, (1958).

6. DAVIES, H. G. and ENGFELDT, B. *Lab. Invest.* **3**, 277 (1954).
7. DAVIES, H. G. and ENGSTROM, A. *Exp. Cell Res.* **7**, 243 (1954).
8. DAVIES, H. G., ENGSTROM, A. and LINDSTROM, B. *Nature, Lond.* **172**, 1041 (1953).
9. DAVIES, H. G., WILKINS, M. H. F., CHAYEN, J. and LA COUR, L. F. *Quart. J. Micr. Sci.* **95**, 271 (1954).
10. DYSON, J. *Proc. Roy. Soc.* A **204**, 170 (1950).
11. DYSON, J. *J. Opt. Soc. Amer.* **47**, 557 (1957).
12. HALE, A. J. *J. Physiol.* **125**, 50P (1954).
13. HALE, A. J. *J. Roy. Phys. Soc. Edinb.* **24**, 44 (1955).
14. HALE, A. J. *Exp. Cell Res.* **10**, 132 (1956).
15. HALE, A. J. "The Interference Microscope in Biological Research." E. & S Livingstone Ltd., Edinburgh, (1958).
16. HALE, A. J. and MORRISON, S. D. unpublished observations.
17. HALE, A. J. and KAY, E. R. M. *J. Biophys. Biochem. Cytol.* **2**, 147 (1956).
18. HUXLEY, A. F. *J. Physiol.* **117**, 52P (1952).
19. HUXLEY, A. F. *J. Physiol.* **125**, 11P (1954).
20. HUXLEY, A. F. *J. Physiol.* **133**, 35P (1956).
21. JOHANSSON, L. P. and AFZELIUS, B. M. *Nature, Lond.* **178**, 137 (1956).
22. LAGERLOF, B., THORELL, B. and AKERMAN, L. *Exp. Cell Res.* **10**, 752 (1956).
23. LEUCHTENBERGER, C., MURMANIS, I., MURMANIS, L., ITO, S. and WEIR, D. R. *Chromosoma* **8**, 73 (1956).
24. MARSDEN, N. V. B. *Exp. Cell Res.* **10**, 755 (1956).
25. MELLORS, R. C. *Tex. Rep. Biol. Med.* **11**, 693 (1953).
26. MELLORS, R. C. and HLINKA, J. *Exp. Cell Res.* **9**, 128 (1955).
27. MELLORS, R. C., KUPFER, A. and HOLLENDER, A. *Cancer* **6**, 372 (1953).
28. MELLORS, R. C., STOHOLSKI, A. and BEYER, H. *Cancer* **7**, 873 (1954).
29. RICHARDS, O. W. *J. Biol. Photogr. Ass.* **19**, 7 (1951).
30. SIRKS, J. L. *Ann. Phys. Chem.* **18**, 458 (1894).
31. SMITH, F. H. *Research* **8**, 385 (1955).

FLYING SPOT MICROSCOPY

W. K. Taylor

Department of Anatomy, University College, London

APPARATUS AND TECHNIQUES FOR OBTAINING TELEVISION PICTURES AND AUTOMATIC SIZE DISCRIMINATING COUNTS OF MICROSCOPIC PARTICLES

Introduction

The flying spot microscope [2] consists of a cathode ray tube (C.R.T.), a microscope and a photomultiplier cell, together with the associated electrical circuits. The arrangement is illustrated in Fig. 1(b), which should be compared with that of the flying spot film scanner (Fig. 1(a)). In both systems the flying spot of the C.R.T. is deflected along a set of parallel lines as in a television receiver "raster" but the brightness of the spot is kept constant. The raster of lines is projected at a reduced size on to the film or microscope slide specimen and the light that passes through the specimen at each instant is converted into a voltage (video signal) by the photomultiplier. If a black and white picture of the specimen is required it is readily obtained by supplying this video signal to a television monitor cathode ray tube. The output voltage of the flying spot scanner or flying spot microscope may also be analyzed by apparatus that will automatically count the number of particles in the specimen and measure their size distribution. The considerable saving of labour that can be achieved by automatic sizing becomes of increasing importance as the number of specimens that have to be analyzed increases, since the manual size analysis of a single specimen may take several days. Experience has shown that there can be a large variable element in manual sizing and counting, especially when the particles are not of uniform opacity. Machine counting is far more consistent, once the controls have been adjusted to suit the opacity of the particles and the background.

An alternative method of obtaining pictures of specimens is shown in Fig. 1(c). A standard industrial television camera is arranged so that the image of the specimen is projected on to the light-sensitive surface of the camera tube. At the time of writing the television microscope has not been used to supply counting and sizing apparatus, the most suitable input system for this application being the flying-spot film scanner. The use of the latter involves the intermediate step of preparing a negative enlargement of the specimen taken at an appropriate magnification by photomicroscopy, using a light, ultraviolet or electron microscope. On a first impression, it seems that the flying-spot microscope is preferable since it eliminates the necessity for preparing negatives. Experience has shown, however, that the numerous adjustments which include slide position, focus, magnification and sensitivity, are difficult to reproduce, especially by an untrained opera-

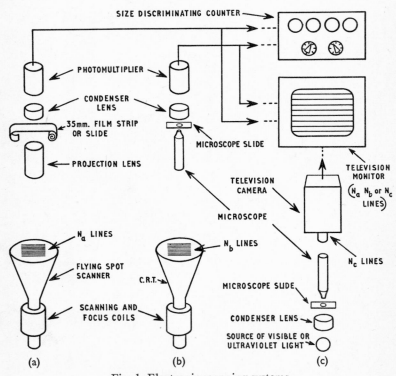

Fig. 1. Electronic scanning systems.

tor. In the flying-spot film scanner, on the other hand, there is only one variable, the sensitivity to light, and even this can be fixed if the negatives have uniform background opacity and if the range of variation of the ratio between the opacity of the particles and background is approximately constant.

Resolving Power of Scanning Systems

It is intuitively obvious that the maximum number of particles that can be counted independently by a scanning system will increase as the number of lines in the raster increases. There is no point in increasing the number of lines beyond a certain limit, however, since the number of independent lines N_{max} that can pass through the microscope objective or alternative lens system is limited by the numerical aperture of the lens α, the wavelength λ of the light emitted by the cathode ray tube and the maximum width of the raster on the specimen, X, according to the equation [1].

$$N_{max} = \frac{2\alpha X}{\lambda}$$

There is no point in increasing the number of raster lines N above this value, since with N_{max} lines there is no loss of information between the lines and also no duplication of information in neighbouring lines. At a mean wavelength of 0.5×10^{-3} mm, a lens of numerical aperture 1.3 will allow approximately 1,000 independent lines to pass when X is 0.2 mm. It is not possible at present, to obtain a cathode ray tube that will give this number of independent lines, the limit being in the region of 500 lines. At very small spot sizes there is a tendency for the brightness to vary as the spot crosses the grains of the phosphor, and this constitutes a further limitation, although it can be corrected to some degree by negative feedback.

When counting and sizing is required, it is usually desirable to work at the lowest magnification for which the smallest particles are crossed by one or two lines, since the number of particles that can be analyzed in one complete scan of N lines is then a maximum. At higher magnifications a proportionately smaller area of specimen is covered by the scan and more films must be prepared for the film scanner, or in the case of the direct microscope method (Fig. 1(b)) the slide must be moved to more positions.

The picture of a specimen obtained on a television monitor is made up of the same number of lines N as the raster on the flying spot scanner, usually spaced farther apart to produce a larger picture. If N could be made equal to N_{max} the picture would contain practically all the information in a photograph taken with an equivalent exposure, but in practice the reduced value of N, together with electrical noise effects lead to inferior pictures. Electrical noise generated in the photomultiplier contributes to poor pictures and inaccurate counting when the light signals are reduced at high magnifications. At the low magnifications employed in a film scanner, however, photomultiplier noise voltage is much smaller than the voltage due to the light fluctuations, and this is a second reason for employing the film method.

The fact that the number of scanning lines is normally less than the maximum number of lines that a lens will allow to pass through without overlap, does not limit the resolving power if one is content to examine a fraction of the specimen area that the lens will cover. If, for example, the lens will pass 1,000 independent lines but only 500 scanning lines are available, they can be concentrated in one half of the specimen at the minimum spacing $\lambda/2\alpha$. If the 500 lines cover a smaller area than this, the limitation of resolving power will be due to the lens system and not the scanning system.

The foregoing limitations are illustrated in Fig. 2 which shows photographs of the television monitor screen. A picture of myelinated

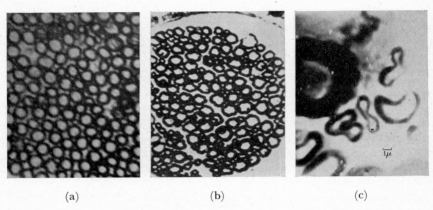

(a)   (b)   (c)

Fig. 2. Myelinated nerve pictures obtained (a) on flying-spot, and (b) (c) on television microscope monitor.

nerve fibers taken with a flying spot microscope giving 405 lines is shown at (a). Pictures (b) and (c) were obtained with a television microscope giving 625 lines. The improvement from (a) to (b) is due partly to the increased number of lines and partly to an improvement in signal/noise ratio. The latter effect would presumably be still more important at ultraviolet wavelengths since it is easier to produce a continuous source of ultraviolet light of suitable wavelength than it is to produce a cathode ray tube that will give an equivalent output. An ultraviolet television microscope system developed in the U.S.A. contains three ultraviolet sources of different wavelength and gives continuous colour pictures of unstained living cells for periods of half an hour or more without killing the cells. This has been made possible by the development of television cameras with high sensitivity to ultraviolet light.

In Fig. 2(c) the magnification is increased by allowing only a small part of the specimen image to be picked up by the camera. The 625 lines are more than adequate to reproduce this section of the image, the limitation of resolution being due to the microscope objective.

Size Discrimination and Counting

Apparatus for operating on a video signal in such a way that the number and size of the scanned particles can be measured automatically, is independent of the scanning system employed and the same apparatus can be supplied, together with a television monitor, by the output of a film scanner, microscope slide scanner or possibly an electron microscope scanner in which a television camera tube surface

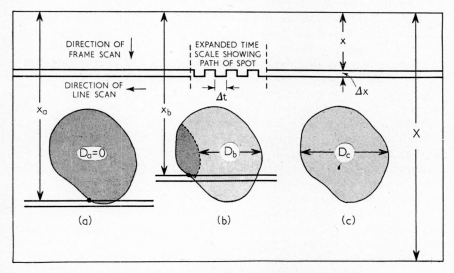

Fig. 3. Illustrating size discrimination and counting.

replaces the fluorescent screen of the conventional electron microscope.

The essential property of accurate counting apparatus is that each particle should be counted only once, irrespective of its size and the number of times that it is crossed by the scanning lines. A very small particle may be crossed only once but a large particle in the same specimen may be crossed ten times or more. The basic method of preventing double counts by means of a double spot is still widely employed, but a number of different methods of obtaining a double spot or its equivalent are in use. The most important of these are the delay method and the spot vibration method. In the former, a line of video signal is stored and recovered during the following line and in the latter, the spot is made to follow a double line path in alternating steps as shown in Fig. 3. The switching method leads to relatively simple and inexpensive equipment and has been adopted in the new apparatus at University College. It is known that a video signal containing no frequencies above F_{max} can be accurately represented by samples spaced in time by an interval not greater than $\frac{1}{2F_{max}}$. If the N lines are produced in T seconds, the time taken for the spot to travel a line is T/N seconds and the time taken for the spot to travel a distance equal to the line separation is T/N^2 seconds. An estimate of F_{max} is N^2/T which for 200 lines and T = 4 seconds is 10,000 cycles/second. The switching interval must be less than 1/20,000 second and can, in fact, be made one tenth of this to simplify the filter circuits.

The way in which the double line eliminates multiple counts is illustrated in Fig. 3(a). In this representation of the raster a single particle is shown at three different settings of the size discriminator. At the moment we shall be concerned only with the particle as shown at (a). The two lines, separated by the distance Δx, are assumed to move from the top to the bottom of the rectangular raster or frame in steps of Δx. A count is made only if the top line of the pair crosses the particle and the bottom line does not cross the particle. The actual instant of counting occurs when the top line is leaving the particle, and since the spot moves from right to left in this diagram the count occurs at the position marked by a black dot. This counting condition happens only once for this particle, as can be verified by moving the two lines upwards or downwards in steps of Δx.

Size discrimination [3] is achieved by means of circuits that modify the video signals obtained from the two lines before they are applied to the counter. The circuits simply delay the edge of the particle that is first encountered by the spot by a variable amount D, which is denoted by D_b in case (b). When D_b is such that the delayed leading edge would fall beyond the lagging edge it is eliminated, with the result that the modified video signal corresponds to a "remainder particle" bounded by the delayed leading edge and the lagging edge and corresponding to the darker area in Fig. 3(b). If the modified video signal is supplied to the counter circuit, a single count is obtained at the point marked by a black spot on the lower end of the dark area. It will now be apparent that when D is equal to or greater than the size of the particle in the direction of the line scan, there will be no modified particle and hence no count. The case in which D is just equal to the particle size is illustrated at Fig. 3(c). During one frame, in which the lines move from x = 0 to x = X in steps of Δx, the value of D is held constant at a preselected value and a count of all particles with a dimension greater than D in the line scan direction is obtained. A size histogram is derived by taking the differences of consecutive counts, obtained as D is increased in steps from zero to the value at which the count is zero. This maximum value of D corresponds to the size of the largest particle in the specimen.

A test slide containing twelve circular particles of four different sizes, large, medium, small and very small, was used to obtain the four monitor pictures shown in Fig. 4. The top row shows the twelve particles, each with a counting spot derived from the counting circuit in the lower left-hand position as illustrated in Fig. 3(a) for the zero value of D. The total count, as indicated on dekatron counters, is 12, since each particle is counted once. In the second row, the value of D has been increased until the very small particles are not counted. The

COUNT NO:	COUNT ANALYSIS				L = LARGE M = MEDIUM S = SMALL VS = VERY SMALL	TOTAL COUNT
	L	M	S	VS		
1	3	3	3	3		12
2	3	3	3	0		9
3	3	3	0	0		6
4	3	0	0	0		3

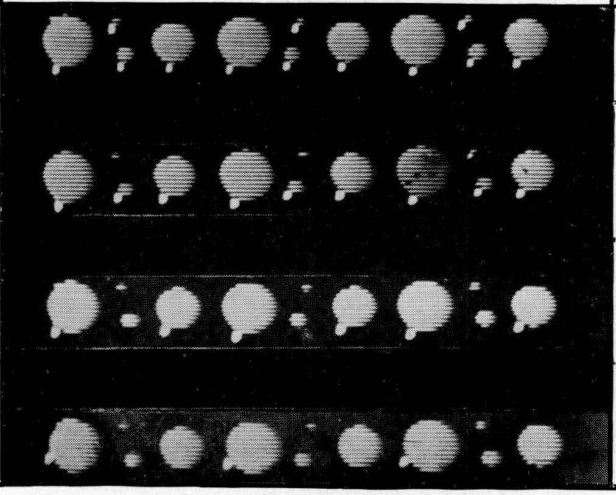

Fig. 4. Size discriminating counts of twelve particles.

total count is thus 9 and is made up of 3 large, 3 medium and 3 small particle counts. The difference between these total counts, i.e. 3 would be the first point on the size histogram. In the third row only the large and medium size particles are counted and on the bottom row the three largest particles remain. A further increase in D makes the total count zero.

A Practical Size Discriminating Counter

A system that has proved reliable in operation and which gave the results shown in Fig. 4, is illustrated in Fig. 5. The scanning system is the type (a) of Fig. 1 in which photographic negatives of the specimens are scanned. The video signal output of the photomultiplier is switched by means of an electronic circuit, of which only the mechanical equivalent is shown, to two low pass filters that smooth the pulses of voltage into continuous waveforms. At this stage a decision has to be made that corresponds to the decision made during a visual count when a particle is only very slightly darker than the background and does not have clearly defined edges. It is clearly necessary at some stage to neglect small changes of opacity that are below an arbitrarily chosen level as being insignificant compared with more clearly defined particles. In this apparatus the decision is under the control of the operator in the form of a clipping level control before the Schmitt trigger which eliminates signals corresponding to particles with less than a given contrast relative to the background. In general, it is found that the

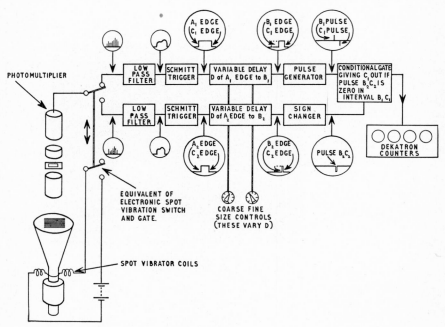

Fig. 5. Size discriminating counter.

clipped signals corresponding to the two lines are slightly different, as illustrated by the waveforms contained in the circles after the Schmitt triggers. The usual convention of time increasing from left to right is adopted and A_1, A_2; C_1, C_2 are therefore leading and trailing edges respectively. Coarse and fine size controls are provided and as explained previously these vary the delay D of the leading edges by equal amounts to B_1 and B_2. A pulse generator delivers positive and negative pulses, at the instants when the edges B_1 and C_1 occur, to a conditional gate. Pulse B_1 opens the gate, which remains open until C_1 occurs if a second pulse B_2C_2 has not closed it during the intervening period. The gate is only open when C_1 occurs if the counting condition is fulfilled, since only then is B_2C_2 absent. When this happens, C_1

passes through the gate to the dekatron counters which increase their indication by unity. A second electronic switch working in synchrony with the first, drives a square wave of current through spot vibration coils round the scanning cathode ray tube. The magnetic field deflects the spot as shown on the expanded time scale in Fig. 3. The complete circuit of this apparatus is contained in a box measuring approximately $2' \times 1\frac{1}{2}' \times 1\frac{1}{2}'$ and is of approximately the same complexity as a television receiver.

Preparation of Films

The apparatus described above and illustrated in Fig. 5 is designed to accept 35 mm negative film in the form of slides or continuous strip. When film strip is used the film can be wound manually one frame at a time, size discriminating counts being made of each frame by selecting appropriate positions of the coarse and fine size controls. In preparing the film with the aid of a light, ultraviolet, or electron microscope, it is important to choose the magnification so that the smallest particles of interest are not smaller than the line separation Δx in the image of the raster as it appears on the film in the scanning apparatus. This is equivalent to the condition that the actual size of the smallest particle on the negative should not be less than $20/N$ mm if the scanned length X (Fig. 3) is 20 mm and if there are N lines in the raster. Thus if N is 200 the smallest particles to be counted on the film should be 0.1 mm and if the magnification is approximately 100 these will correspond to 1 μ particles in the actual specimen. In some applications the actual particles may be larger than the minimum size and in this event they can be reduced in size on the film. The accuracy with which particle sizes can be measured is limited to $\pm$ a length that is of the order Δx on the film. The percentage error will thus be quite large for the smallest particles and will decrease steadily as the particle size increases, a particle of size $10.\Delta x$ being measured with an average accuracy of $\pm 10\%$. For high accuracy it is therefore necessary to choose a magnification that will make the particles to be measured accurately cover 10 or more lines of the raster on the film. In practice, the overall results for a large number of particles tend to be more constant than the results for a single particle, since the positive and negative errors tend to cancel to give an average figure. Visual counting is subject to similar errors when the particle size is estimated to the nearest division on a scale.

Other errors common to automatic and visual counting originate from the continuous nature of opacity in biological specimens. Changes in opacity from one particle to the next can be due to variations in density, thickness or distance from the plane of focus. The visual counter is always faced with the arbitrary decision of whether or not to count a fuzzy area of slightly greater opacity than the background as a particle. If he decides to include the fuzzy area, he is faced with the second problem of measuring its size. This again is an arbitrary choice if the opacity of the area gradually declines from a maximum to the opacity of the background. He could estimate the half-way point by eye, but it is clear that more consistent results could be expected if he measured the maximum opacity by letting the light from this point fall through a small hole on to a photomultiplier. By moving the small hole along the contour for which the photomultiplier reading corresponds to an opacity half-way between the maximum and the background, the "size" of the particle could be estimated. This is effectively what the machine does automatically.

The foregoing decisions are easily made if the opacity of the particles is large compared with that of the background. This ideal condition can be approached in reverse by taking advantage of the increase in contrast given by the photographic process. On the negative film the particles appear as translucent areas and the contrast between these areas and the darker background can be much greater than that of the specimen. It is, however, important to standardize the contrast if comparable results are to be obtained from films of different specimens. This is because all particles producing more than a threshold translucency on the film are counted. A slight decrease in translucency may therefore eliminate particles that were just above threshold, although it may have no effect if all the particles were well above threshold. It is therefore important to take every advantage of staining techniques, etc., that will give good contrast in the specimen.

EXAMPLES OF AUTOMATIC COUNTING AND SIZING

A microphotograph of a section of rat brain, stained to show the cells but not their processes, is shown in Fig. 6(a). The negative of this picture on 35 mm film measures 34 × 22 mm and the scanner is

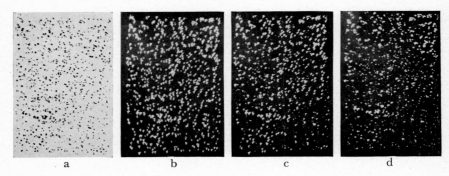

Fig. 6. Rat brain cells (a) and size discriminating counts (b) (c) and (d).

adjusted to cover this area. Five total counts of the cells, obtained by setting D to zero, are indicated in Fig. 7 by the full lines between zero and 1 μ. It will be noticed that the five counts are all with ± 5% of the mean whereas the differences between three visual counts, denoted by dash lines, are much greater. By displaying the particles on the television monitor and superimposing a white dot, corresponding to the black dot in Fig. 3(a), to indicate counts, the composite picture of Fig. 6(b) is obtained. If the size controls are now advanced to increase D

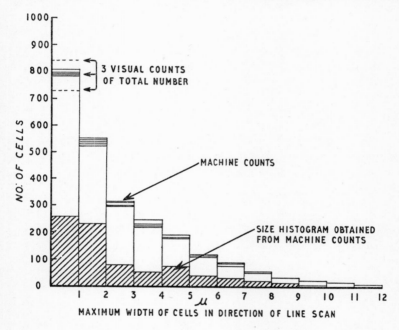

Fig. 7. Size histogram of rat brain cell sample.

the white dots are gradually eliminated, firstly from the smallest and then from the larger particles as shown at Fig. 6(c) and (d). The actual results as indicated on the dekatron counter are plotted in Fig. 7, for a range of D values corresponding to cell sizes from 0 to 12 μ in steps of 1 μ. There are five counts for each setting but in some cases two or more of these are identical. A histogram, obtained by taking differences of consecutive average counts, is indicated by the shaded area.

The interpretation of the histogram in histological terms is not obvious in this example, since neurones and neuroglia both appear in unknown proportions and the nucleoli of some cells are so much more opaque than the cytoplasm that they may be counted when the cytoplasm is missed through being out of focus or poorly stained. The preparation of specimens in the form that is suitable for the apparatus is clearly a subject that will require considerable attention before the automatic technique can be extended to the counting and sizing of complex biological material.

REFERENCES

1. FRANCIA, G. Toraldo di. *J. Opt. Soc. Amer.* **45**, 7, 497 (1955).
2. ROBERTS, F. and YOUNG, J. Z. *Proc. Inst. Elect. Engrs* **99**, Part 111A, 20, 747 (1952).
3. TAYLOR, W. K. The Physics of Particle Size Analysis. *Brit. J. Appl. Phys.* Suppl. No. 3 (1954).

AUTHOR INDEX

Numbers in parentheses are reference numbers and are included to assist in locating references when the authors' names are not mentioned in the text. Numbers in italics refer to the page on which the reference is listed.

A

Abbot, J., 81 (46a), *87*
Ada, G. L., 70 (26b), *86*
Afzelius, B. M., 173 (21), 176 (21), *186*
Akerman, L., 185 (22), *186*
Alexander, W. R. M., 74, 78 (1, 2), *86*
Allison, A. C., 79 (69a), *88*
Anderson, N. G., 90, *94*
Andrews, A. M., 182 (1), *185*
Avery, O. T., 161, *171*

B

Balfour, B., 79 (69a), *88*
Ball, G. H., 111, *116*, 121
Baltzer, F., 110, 114, *117*, 122
Barbu, E., 109, *118*
Bardawil, W. A., 79 (2a), *86*
Barer, R., 178 (2), 182 (3), 184 (3), *185*
Barnard, E. A., 134 (1), *137*, 147 (6), 149 (9), *153*
Barrett, F. C., 96 (9), *117*
Barter, R., 151 (12), *153*, *185*
Battaglia, E., *65*
Baud, CH.-A., 64, *65*
Beale, G. H., 69 (3), 75, *86*
Bellairs, R., 48, *57*
Beneš, L., 64, *66*
Berenbaum, M. C., 68 (4, 4a), 70 (4), *86*
Berg, B., 104 (81), *118*
Bergel, F., 142 (5), *143*
Berliner, E., 67 (18), *86*
Bernecky, J., 72 (43a), *87*
Bernstein, R. E., 127 (2), *137*
Berridge, N. J., 109 (2, 15), *116*, *117*
Berry, H. K., 108 (3, 4, 5, 47), *117*, *118*
Berton, W. M., 108 (102), *119*
Beyer, F. G., 110, *118*, 122, 123
Beyer, H., 184 (28), *186*
Bing, J., *86*
Birnbaum, D., 74, *86*
Block, J. R., 96 (6, 7), *117*
Boddy, R. G. H. B., 156 (10), *171*
Bohme, H., 106, *119*
Boivin, A., 160, *171*
Boothroyd, E. R., 163 (3), *171*

Boss, J., 60 (6), 61 (5, 6), 62 (6), 64 (4), *65*
Bottazzi, V., 109 (2, 8, 15), *116*, *117*
Boursnell, J. L., 70, *86*
Bowyer, F., 133 (3), *137*
Boyer, G. S., 79 (6a), *86*, 169 (4, 21), *171*
Brachet, J., 84 (7), *86*, 156, 160 (5), *171*
Bragg, A. N., 63, *65*
Brandt, P. W., 85, *86*
Briggs, R., 2, 4, 5 (4, 15), 7, 10 (1, 3, 13), *14*
Brimley, R. C., 96 (9), *117*
Brncic, D., 103, *118*
Buetner, E. H., 75, 76, *86*
Bush, V., 69 (10), *86*
Butzel, H. M., 106, *117*
Buzzati-Traverso, A. A., 97 (12), 99 (12), 102 (11, 12), 106, 107, 108, 111 (12, 13), 113 (13), *117*, 123

C

Cain, L., 108 (3, 4), *117*
Callan, H. G., 25 (1), 27 (1, 2), 29 (1), 30 (4), 31 (1), 33 (1), 35 (1), 40 (5), 42 (3), *45*, *46*, 64, *65*, 66
Carver, R. K., *87*
Caspersson, T., 146 (2), *153*, 156, 157, 160 (7), 165, 168 (7), *171*
Cassidy, H. G., 96 (14), *117*
Castiglioni, M. C., 104, *118*
Cavanaugh, M. C., 81 (46a), *87*
Chadwick, C. S., 68 (11), 69 (64), 70, 73 (64a), 77 (64), *86*, *88*
Chambers, R., 65, *66*
Chargaff, E., 156, *171*
Chayen, J., 64 (35, 36), *66*, 181 (9), 184 (9), *186*
Cheeseman, G. C., 109 (2, 15), *116*, *117*
Chen, P. S., 101 (16, 18), 104, 110, 114, *117*, 122
Christensen, H. N., 131, *137*
Clark, E. W., 111, *116*, 121
Clark, P. J., 108 (108), *119*
Clarke, S., *86*
Clarke, W. M., 83, *88*

Claugher, D., 110 (120), *119*
Clayton, R. M., 68 (12, 13), 69 (12, 13, 14), *70* (12, 13), 74 (16a), 76 (15), 81, 84 (14, 16, 84), *86*, *88*
Cleveland, L. R., 62 (13, 14), 63, *66*
Cohen, S., 79 (66a), *88*
Commandon, J., 2, *14*, 15, *22*
Conn, H. J., 70, *86*
Connolly, J. M., 77 (21, 54, 94), *86*, *87*, *88*
Consden, R., 95, *117*
Conway, E. J., 126 (5), *137*
Coombs, R. R. A., 70 (6), *86*
Coons, A. H., 67, 68 (22), 69, 70, 77 (20, 21, 54, 94), 78, 79, *86*, *87*, *88*, 148, *153*
Cowgill, R. W., 92, *94*
Craig, J. M., 78 (36), *87*
Cramer, F., 96 (20), *117*
Creech, H. J., 67 (18), *86*
Cruickshank, B., 68, 69 (42), 72, 75, 79, 82 (26), *86*, *87*
Crumpler, H. R., 108 (21), *117*
Cummins, C. S., 109, *117*
Curran, P. F., 133 (34), *138*
Curry, A. R. G., 75, *86*
Curtain, C. C., 70 (26a), *86*

D

Dalgaard, J. B., 152 (13), *153*
Danielli, J. F., 2, *14*, *22*, 128 (7), 129 (36), 135, *137*, *138*, 140 (2), 141 (3, 4), *143*, 147 (4, 5, 6), 151 (12), *153*, 156, *171*, *185*
Danneel, R., 51, *57*, 101 (23, 24), *117*
Dannevig, E. H., 113, *117*, 123
Danon, M., 25 (15), *46*
Darlington, C. D., 38 (6), *46*, 63, *66*
Datta, S. P., 113, *117*, 123
Davidson, J. N., 156, *171*
Davies, H. G., 151 (12), *153*, 156 (10), *171*, 173 (5), 179 (5), 181 (8, 9), 184 (9), 185 (6, 7), *185*, *186*
Davis, A. M., 163 (25), 166 (25), *171*
Davson, H., 128 (7), *137*
de Fonbrune, P., 2, *14*, 15, *22*
De Grouchy, J., 99 (28), 108 (28), *117*
De Lerma, B., 99 (30, 31), *117*
De Marsh, Q. B., 64, *66*
Denny, F. W., 79 (6a), *86*
Dent, C. E., 107, 108 (21), *117*
Detre, K., 78 (33), *86*
De Vincentiis, M., 99 (30, 31), *117*

Dick, D. A. T., 182 (3), 184 (3), *185*
Dineen, J. K., 70 (26b), 74 (47), *86*, *87*
Dixon, E. J., 79 (83a), *88*
Dobzhansky, Th., 108 (5, 47), *117*, *118*
Dodson, E. O., 25 (7), *46*
Doniach, I., 76 (28), *86*
Dupont-Raabe, M., *117*
Durrum, E. L., 96 (6), *117*
Duryee, W. R., 23 (8), 25 (9), 28, *46*
Dyson, J., 173, 176 (10, 11), 182 (10), *186*

E

Ebert, J. D., 70 (30), 80, *86*
Edsall, J. T., 127 (8), *137*
Egelhaaf, A., 99 (112, 113), 100 (113), 104 (37, 59, 82), *117*, *118*, *119*
Ellis, J. P., 110 (89), *118*, 120, 121
Elsdale, T. R., 8, 9 (9), 12 (11), *14*
Emmart, W. E., 79, *87*
Engfeldt, B., 185 (6), *186*
Engle, K., 70 (43), 76 (43), 84 (43), *87*
Engstrom, A., 181 (8), 185 (7), *186*
Eschrich, B., 101 (23), *117*

F

Feldman, M., 69 (13), 70 (13), 81, *86*
Fell, H. B., 60, 65, *66*
Fernandez-Moran, H., 51, *57*
Finch, S., 78 (33), *86*
Finck, H., 68 (31), 69 (31), 73 (46), 79, 80 (46), 81 (57a), *86*, *87*
Finley, H. E., 106, *117*
Fishberg, M., 8, 9 (9), 12 (11), *14*
Flemming, W., 61, 63, *66*
Flickinger, R. A., 84, *86*, *88*
Foote, F. W., 70 (71), 72 (71), *88*
Forrest, H. S., 99 (39, 40, 41, 42), 100, 103, *117*, 121 (99)
Fox, A. S., 99, 103, 115, *117*, *118*, 121
Francia, G. Toraldo di, 189 (1), *198*
Francis, M. D., 84 (32), *86*
Fraser, K. B., 73 (64a), *88*
Friou, G. S., 78 (33, 34, 35), *86*, *87*
Frolova, S. L., 64 (18), *66*
Fujito, S., 106, *118*

G

Gahan, P. S., 64, *66*
Gall, J. G., 23 (12), 25 (11, 12), 29 (14), 30 (14), 31 (13), 33 (12), 38, *46*

Gárdos, G., 127 (43), *138*
Gartler, S. M., 108 (5, 47), *117*, *118*
Gay, H., 64 (28), *66*
Gentile, D. P., 169 (32), *172*
Gerbasi, J. R., 75 (9), 76 (9), *86*
Gibson, F. J., 110 (90), *118*, 119, 120, 121, 122
Ginsberg, H. S., 79 (6a, 72), *86*, *88*, 169, *171*
Gitlin, D., 76, 78 (36), *87*
Glynn, I. M., 129 (9), 130 (9), 131 (10), *137*
Glynn, L. E., 69 (38), 73, 74, *87*
Goldacre, R. J., 135, *137*
Goldman, M., *87*
Goldschmidt, E., 103, *118*
Goldstein, D., 75 (43b), *87*
Goldwasser, R. A., 70, *87*
Gonnard, P., 111, *119*, 120
Goodman, J., 132 (12), *137*
Goodman, M., 72, 73, *87*
Gordon, A. A., 95, *117*
Graf, G. E., 103, *118*
Grassé, P.-P., 62 (19), *66*
Greenspon, S. A., 72, 73 (40, 50a), *87*
Gregory, M., 109 (50, 51), *118*
Gregory, P. W., 158 (31), 161 (31), 163 (31), 169 (31), *172*
Griffiths, L. A., 106 (52), *118*
Gross, J., 76 (53), *87*
Gurdon, J. B., 8, 12, *14*
Guyénot, E., 25 (15), *46*

H

Hadorn, E., 97, 99 (53, 54, 55, 57, 111), 100, 101 (18, 56, 57, 61, 65, 122), 103, 104 (59, 60), 105, *117*, *118*, *119*
Haldane, J. B. S., 116, *118*
Hale, A. J., 151 (11), *153*, 173 (15), 179 (15), 181 (17), 182 (1, 13, 17), 184 (12, 16), 185 (14), *185*, *186*
Harris, E. J., 126 (14), 130 (13, 14), *137*
Harris, H., 107 (35), 108, 109, 113, *117*, *118*, 123
Harris, R. H., 110 (120), *119*
Hasborne, J. B., 106 (68), *118*
Hatton, E., 84 (31a), *86*
Hebborn, P., 141 (3), *143*
Heberer, G., 61 (20), *66*
Helweg-Larsen, H. F., 163 (22), *171*
Henke, K., 54, *57*
Hertig, A. T., 79 (2a), *86*

Hertwig, G., 160, *171*
Hess, E. L., 77 (79), *88*
Hewitt, R. E., 69 (10), *86*
Hill, A. G. S., 69 (42), 72, *86*, *87*
Hill, R. F., 70 (71), 72 (71), 79, *88*
Hiramoto, R., 70, 72 (43a), 75, 76, 84, *87*
Hlinka, J., 184 (26), *186*
Hochman, B., 99, *118*
Hoenigsberg, H. F., 104, *118*
Holborow, E. J., 69 (38), 73 (38, 45), 74, 78 (44), *87*
Holden, J. T., 99, *118*
Hollender, A., 182 (27), 184 (27), *186*
Holman, H. R., *88*
Holt, S. J., 148, *153*
Holter, H., 85 (57b), *87*
Holtzer, H., 68 (31), 69 (31), 73 (46), 79, 80, 81 (46a, 46b, 57a), 85, *86*, *87*
Horvath, B., 79, *87*
Hsu, T. C., 163 (12), *171*
Huf, E. G., 126, *137*
Hughes, A. F. W., 60, *66*
Hughes, P. E., 74 (47, 47a, 48a), 75 48a), *87*
Hughes-Schrader, S., 158 (31), 161 (31), 163 (31), 169 (31), *172*
Huxley, A. F., 182, *186*

I

Ito, S., 161 (28), 163 (28), 170 (13, 28), *171*, 184 (23), *186*

J

Jankovic, B. D., 74, *87*
Jacob, J., 10 (25), *14*
Jacobson, W., 64 (23), *66*
Janeway, C. A., 78 (36), *87*
Janssens, F. A., 61 (24), *66*
Jérome, M., 111, *119*, 120
Johansson, L. P., 173, 176, *186*
Johnson, G. D., 69 (38), 73, 78 (44), *87*
Jolly, P. C., 129 (25), 130 (25), *137*
Jones, N. R., 113, *118*
Jones, R. N., 67 (18), *86*
Jurandowski, J., 72 (43a), *87*

K

Kacser, H., 69 (3), 75, *86*
Kaplan, M. H., 67 (19), 69, 79, *86*, *87*
Kaplan, W. D., 99, *118*

Karrer, P., 99 (111, 114, 115, 116, 117), 100 (111), *119*
Kater, J. McA., 60, *66*
Kaufmann, B. P., 60, 64 (28), *66*
Kautz, J., 64, *66*
Kay, E. R. M., 181 (17), 182 (17), *186*
Kikkawa, H., 104, 106, *118*
King, E. S. J., 74 (48a), 75 (48a), *87*
King, T., 2, 4, 5 (4, 15), 7, 10 (1, 3, 13), *14*
Kinosita, R., 61 (41), *66*
Kirk, R. L., 110, *118*, 122, 123
Klatzo, I., 79, *87*
Klein, E., 166 (23), *171*
Klein, G., 166 (23), *171*
Knowles, M. F., *117*
Koller, I., 166, *171*
Koller, P. C., 63, *66*
Koref, S., 103, *118*
Kossel, A., 155, *171*
Krakower, C. A., 72, 73 (40, 50a), *87*
Kühn, A., 99 (112), 100 (60), 104 (60, 80, 81, 82), *118*, *119*
Kupfer, A., 182 (27), 184 (27), *186*
Kurosumi, K., 60, 64, *66*
Kursteiner, R., 100 (61), 101 (61), *118*
Kuwada, Y., 65, *66*

L

La Cour, L. F., 64 (35, 36), *66*, 181 (9), 184 (9), *186*
Lacy, P. E., 76, *87*
Lagerlof, B., 185 (22), *186*
Landing, B. H., 76 (37), *87*
Langenbek, W., 106, *119*
Langman, J., 84 (51), *87*
Laufer, H., 81, *87*
Laven, H., 104, *118*
Leblond, C. P., 76 (53), *87*
Lebrun, J., 79, *87*
Lederberg, J., 78, *88*
Leduc, E. H., 77 (21), 77 (54), *86*, *87*
Lee, J. W., 110, 113, *118*, 122
Lee, K. Y., 109, *118*
Lehman, H. G., 10 (17, 18), *14*
Lehmann, F. E., 48, *57*
Le Strange, R., 96 (7), *117*
Leuchtenberger, C., 151 (10), *153*, 157 (20), 158 (20, 29, 31, 33), 159, 161 (26, 28, 31), 163 (16, 20, 22, 25, 28, 31, 40), 165 (19, 20, 24, 26, 29, 45), 166 (17, 18, 23, 25, 40), 168 (18), 169 (4, 21, 30, 31, 32, 34, 35, 36), 170 (13, 26, 28), *171*, *172*, 184 (23), *186*

Leuchtenberger, R., 161 (26), 163 (25), 165 (24, 26), 166 (25), 169 (35), 170 (26), *171*
Levene, H., 108 (5), *117*
Lewis, W. H., 65, *66*
Lincoln, T. L., 74, *87*
Lindstrom, B., 181 (8), *186*
Lionetti, F., 133 (34), *138*
Lloyd, L., 42 (3), *45*
Loeser, E., 99 (113, 114, 115, 116, 117), 100 (113), *119*
Lorch, I. J., 2 (8, 19, 20, 21), *14*
Louis, C. J., 74 (47, 48a, 55, 55a, 55b), 75 (48a), *87*
Lund, H., 169 (36), *172*

M

Mabbit, L. A., 109 (50, 51), *118*
McCarty, M., 161 (1), *171*
McDonald, M., 64 (28), *66*
McEntegart, M. G., 68 (11), 69 (64), 70, 77 (64), *86*, *88*
Macgregor, H. C., 30 (4), *46*
Macleod, C. M., 161 (1), *171*
McMahon, P. P., 99 (91), 110 (91), *118*, 122, 123
McMaster, —, 77 (58), *87*
Main, A. R., 110, *118*, 122, 123
Maizels, M., 126 (16), 127 (18, 19), 129 (17), 130, 135 (17), *137*
Marsden, N. V. B., 185 (24), *186*
Marshall, J. M., Jr., 67, 68 (31), 69 (31), 73 (46), 75, 76, 79, 80 (46), 81, 85 (57b), *86*, *87*
Martin, A. J. P., 95, *117*
Martin, W. B., 106, *117*
Mattick, A. T. R., 109 (2, 15), *116*, *117*
Mayersbach, H., 70 (57d), 75 (57d), *87*
Mazia, D., 146 (1), *153*
Meier, R. C., 128 (31), *137*
Mellors, R. C., 72, 78, 79, *88*, 182 (27), 184 (25, 26, 27, 28), *186*
Merker, E., 111, *118*
Micks, D. W., 110, *118*, 119, 120, 121, 122
Miescher, F., 155, *172*
Mitchell, H. K., 97, 99 (39, 40, 41, 42), 100 (39, 40, 41, 62), *117*, *118*
Mitchell, P., 128 (20, 22), 129, 135 (21), *137*
Moore, B. C., 60, *66*
Moore, J. A., 10 (22), 11 (23), 12 (23, 24), *14*

Morgan, T. H., 161, *172*
Morrison, S. D., 184 (16), *186*
Moscona, A. A., 81, *88*
Moyle, J., 128 (22), 129, *137*
Mulherkar, L., 84 (84), *88*
Murmanis, I., 161 (28), 163 (28), 170 (28), *171*, 184 (23), *186*
Murmanis, L., 161 (28), 163 (22, 28), 169 (34), 170 (28), *171, 172*, 184 (23), *186*

N

Nace, G. W., 83, 84, *88*
Nairn, R. C., 68 (11), 69 (64), 70, 73, 77, *86, 88*
Nakamura, T., 65, *66*
Nawa, S., *118*
Nebel, B. R., 60, *66*
Neu, R., 106 (93), *118*
Neuhoff, E., 106 (93), *118*
Nishihara, T., 76 (77), 77, *88*
Nolan, M. D., 99 (91), 110 (91), *118*, 122, 123
Nossal, G. J. V., 78, *88*

O

Ogita, Z., 106, *118*
Ohno, S., 61 (41), *66*
Ohta, G., 79 (66a), *88*
Ord, M. J., 2 (8), *14*
Orlans, E., 79, *88*
Ortega, L. G., 72, 78, *88*
Osborne, R. H., 108 (5), *117*
Osterberg, H., 62 (42), *66*

P

Pantelouris, E. M., 10 (25, 29), *14*
Pappas, G. D., 65, *66*
Pardee, A. B., 92, *94*
Patau, K., 163 (39), *172*
Patlak, C. S., 134, *137*
Pearse, A. G. E., 69 (69), *88*, 152 (15), *153*
Pelc, S. R., 76 (28), *86*
Pepe, F., 81 (57a), *87*
Pereira, H. G., 79 (69a), *88*
Perry, M. M., 48, *57*
Perskey, L., 163 (40), 166 (40), *172*
Pilkington, R. W., 56, *57*
Pinney, E., 61 (44), *66*
Pollister, A. W., 156 (41), 157 (42), *172*

Polonowski, M., 111, *119*, 120
Pomerat, C. M., 163 (12), *171*
Porter, K. R., 2, *14*
Post, R. L., 129 (25), 130 (24, 25), *137*
Pressman, D., 70 (43, 70, 71), 72 (43a, 59, 70, 71), 73, 75 (43b, 43c), 76 (43), 84 (43), *87, 88*
Prince, A. M., 79, *88*
Prosser, C. L., 125 (26), *137*

R

Rasmussen, I. E., 104, 111, 112, 113, *119*, 121, 122
Rechnitzer, A. B., 108, 111 (98), 113 (13, 98), *117, 119*, 123
Redetzki, H. M., 70, *88*
Remington, M., 127 (19), *137*
Rene, A. A., 113, *118*, 122
Reuter, E., 63, *66*
Rhoades, M. M., 38 (16), *46*
Richards, A., 62, *66*
Richards, O. W., 182 (29), *186*
Riggs, T. R., 131 (4), *137*
Ris, H., 28, *46*, 157 (42), *172*
Rizk, V., 70 (6), *86*
Roberts, F., 187 (2), *198*
Robertson, F. W., 103, *119*, 121
Robertson, J. G., 104 (100), 111, *119*, 120
Robertson, J. S., 127 (38), *138*
Roddy, L. R., 113, *118*, 122
Roels, H., 164, *172*
Rogers, L. L., 108 (3, 4), *117*
Rogers, S., 108, *119*
Romanovsky, A., 69 (14), 84 (14), *86*
Rose, G. A., 107 (36), *117*
Rose, N. R., 75 (9), 76 (9), *86*
Rosenberg, T., 134 (27), *137*
Rosenfield, R., 79 (66a), *88*
Ross, R. W., 79, *88*
Ross, W. C. J., *143*
Rothbard, S., 72 (85), *88*
Rothstein, A., 128 (29, 31), 132, *137*
Rounds, D. E., 84, *86, 88*

S

Sambuichi, H., 8, 11, *14*
Sanger, F., 93, *94*
Scharff, T. G., 128 (31), *137*
Schayer, R. W., 77 (79), *88*
Schechtman, A. M., 76 (77), 77, 84 (76), *88*
Scherratt, H. S. A., 106 (68), *118*

Schiller, A. A., 77, *88*
Schoeller, M., 99 (117), *119*
Schrader, F., 158 (29, 31), 159, 161 (26, 31), 163 (31), 165 (26, 29, 45), 169 (30, 31, 32, 34, 35), 170 (26), *171*, *172*
Schütte, H. R., 106, *119*
Schulman, J. H., 128, 136, *137*
Schwinck, I., 100 (63, 64), *118*
Scossiroli, R. E., 113, *119*, 121, 122
Scott, D. G., 73, *88*
Selman, G. G., 48, *57*
Serra, J. A., 63, *66*
Sharp, L. W., 60, *66*
Sharpe, M. E., 109 (15), *117*
Shaw, T. I., 130, *138*
Shepard, C. C., 108 (106), 109, *119*
Shephard, C. C., 70, *87*
Shugar, D., 152 (14), *153*
Sierakowska, H., 152 (14), *153*
Siger, E. T., 79 (66a), *88*
Silverstein, A. M., 68 (80), 70, *88*
Singer, S. J., 70, *88*
Sirks, J. L., 173, *186*
Smith, B. G., 61 (51), *66*
Smith, F. H., 173, 176, *186*
Smith, S., 9 (9), *14*
Soloman, A. K., 133, *138*
Soudek, D., 64, *66*
Spector, W. G., 74 (47), *87*
Spiegel, M., 72 (59), *88*
Spurway, H., 40 (5), *46*
Stein, W. D., 129 (36), 133 (35), 134 (1), *137*, *138*, 149 (9), *153*
Stock, C. C., 142 (5), *143*
Stoholski, A., 184 (28), *186*
Stone, L. S., 83 (81), *88*
Straub, F., 127 (37), *138*
Strauss, L., 79 (66a), *88*
Stumm-Zöllinger, E., 101 (65, 109), *118*
Sutton, H. E., 99 (28), 108 (28, 108), *117*
Swift, H., 163 (39, 46), 165, *172*
Szenberg, A., 152 (14), *153*

T

Taira, T., *118*
Tanaka, T., 163 (48), 164, *172*
Taylor, W. K., 192 (3), *198*
Thorell, B., 146 (3), *153*, 185 (22), *186*
Timonen, S., 163 (49), *172*

Tobie, J. E., 68 (82), *88*
Tomlin, S. G., 25 (20), *46*, 64, *66*
Tosteson, D. C., 127 (38), *138*
Toy, B. C., 79 (2a), *86*
Truscoe, R., 127 (19), *137*
Tuppy, H., 93, *94*
Turba, F., 96 (110), *119*
Tyler, A., 70 (83), *88*

U

Ussing, H. H., 126, 136, *138*

V

Vainio, T., 84, *88*
Vanamee, P., 72 (85), *88*
Van Doorenmaalen, W. J., 69, 81, 83, *86*
Van Duyn, C., Jr., 62 (53), *66*
Vazquez, J. J., 79 (83a), *88*
Vendrely, C., 158 (33), 160, 163 (50), *171*, *172*
Vendrely, R., 158 (33), 160, 163 (50), *171*, *172*
Vincent, W. S., 89, 90, *94*
Viscontini, M., 99 (111, 112, 113, 114, 115, 116, 117), 100, *119*
von Brant, T., 99 (91), 110 (91), *118*, 122, 123

W

Waddington, C. H., 10 (29), *14*, 48, *57*, 84 (84), *88*
Wahl, R., 109, *118*
Walker, B. E., 163 (3), *171*
Ward, J. P., 61 (41), *66*
Watson, M. L., 65, *66*
Watson, R. F., 72 (85), *88*
Webb, M., 64 (23), *66*
Weber, G., 70, *88*
Weiler, E., 73, 74, 85, *88*
Weir, D. M., 78 (44), *87*
Weir, D. R., 161 (26, 28), 163 (28), 165 (26), 169 (32, 34, 35), 170 (26, 28), *171*, *172*, 184 (23), *186*
Weiss, P., *57*
Wenrich, D. H., 61 (55), *66*
Westall, R. G., 108 (21), *117*
Whipple, A., 76 (37), *87*
White, M. J. D., 61 (56), 62 (56), *66*
White, R. G., 76, 77, 78, 84, *88*
Whittam, R., 127 (41), *138*

Wiberg, B., *86*
Wilbrandt, W., 131, *138*
Wilkins, M. H. F., 156 (10), *171*, 181 (9), 184 (9), *186*
Williams, H. B., 106, *117*
Williams, R. J., 108 (118, 119), *119*
Wilson, E. B., 60, *66*
Wilson, E. G., 2 (8), *14*
Winnick, T., 84 (32), *86*
Witebsky, E., 75 (9), 76 (9), *86*
Woerdeman, M. W., 70 (95), *88*
Wolman, M., 74, *86*
Wright, C. A., 110 (120), *119*

Y

Yagi, Y., 73, 75 (43c), *87*, *88*
Yamada, T., 48, *57*
Young, J. Z., 187 (2), *198*

Z

Zerahn, K., 126 (40), *138*
Zeutzschel, B., 51, *57*
Ziegler-Günder, I., 100 (66), 101 (122), 104, *118*
Zimmermann, B., 101, *117*
Zweig, G., 96 (6, 7), *117*

SUBJECT INDEX

A

Active transport, 125
Adenosine triphosphate, 127
Amphibia, 1, 23
Amoeba, 2, 15
Ascites tumour cells, 131
Automatic counting, 196

B

Bacteria, 129
Brain, 75

C

Centrifugation, 90
Centromeres, 38
Chromosomal matrix, 63
Chromosomal vesicle, 60
Chromosomes, 13, 23, 59, 89, 163
Chemotherapy, 139
Cytochemistry, 145, 155
Cytoplasm 15, 47,
Cytoplasmic inheritance, 20

D

Desoxyribonuclease, 29
Differentiation, 47, 67, 70
DNA, 12, 23, 30, 47, 59, 79, 156
Drosophila, 48, 97
Drugs, 139

E

Electron microscopy, 48
Embryonic cells, 1
Embryonic induction, 84
Enzyme, 142, 148
Ephestia, 104
Erythrocytes, 74
Evolution, 19

F

Feulgen stain, 25, 157
Fluorescence, 67
Flying spot microscopy, 187
Freeze drying, 150

G

Gene products, 25, 28, 35, 89
Gut, 73

H

Histidine, 149
Hormones, 75
Human inheritance, 106

I

Interference microscopy, 151, 173
Isotopes, 83, 152

K

Kidney, 72

L

Lampbrush chromosomes, 23
Lens, 81
Liver, 74

M

Mass determination, 173
Microorganisms, 109
Microspectrophotometry, 151, 156
Mitosis, 59
Muscle, 79

N

Nuclear transfer, 1, 15
Nuclear membrane, 64
Nucleic acids, 155
Nucleolus, 9, 59, 89
Nucleoplasm, 63
Nucleus, 1, 15, 50, 59, 89

O

Ommatidium, 48
Oocytes, 24

P

Paper chromatography, 93, 95
Paramecium, 75
Proteins, 141, 145
Protozoa, 106

R

Rana, 2
Ribonuclease, 29
RNA, 47, 93, 156

**THE LIBRARY
UNIVERSITY OF CALIFORNIA
San Francisco Medical Center**

THIS BOOK IS DUE ON THE LAST DATE STAMPED BELOW

Books not returned on time are subject to fines according to the Library Lending Code.

Books not in demand may be renewed if application is made before expiration of loan period.

14 DAY

MAR 23 1969
RETURNED
MAR 12 1969

14 DAY

APR 1 1969
RETURNED
MAR 25 1969
14 DAY

DEC 21 1970
RETURNED
DEC 21 1970

25m-10,'67(H5525s4)4128